Linux Porogramming

LINUX POROGRAMMING

Omkar R. Ghatage
(Ph.D.,M.sc,B.Ed,M.Sc,B.Sc)

NAMYA PRESS.com

Published by

Namya Press

213, Vardhan house, 7/28, Ansari Road, Darya Ganj, Delhi - 110002.

Email: namyapress@gmail.com

Edition : **First Published in 2021 (Namya Press)**

Title : **Linux Porogramming**

Author : **Omkar R. Ghatage**

ISBN : **978-93-91010-05-8**

LINUX PROGRAMMING

UNIT I

Linux Utilities-File handling utilities, Security by file permissions, Process utilities, Disk utilities, Networking commands, Filters, Text processing utilities and Backup utilities, sed – scripts, operation, addresses, commands, applications, awk – execution, fields and records, scripts, operation, patterns, actions, functions, using system commands in awk.

UNIT II

Working with the Bourne again shell(bash): Introduction, shell responsibilities, pipes and input Redirection, output redirection, here documents, running a shell script, the shell as a programming language, shell meta characters, file name substitution, shell variables, command substitution, shell commands, the environment, quoting, test command, control structures, arithmetic in shell, shell script examples, interrupt processing, functions, debugging shell scripts.

Linux Files: File Concept, File System Structure, Inodes, File types, The standard I/O (fopen, fclose, fflush, fseek, fgetc, getc, getchar, fputc, putc, putchar, fgets, gets etc.), formatted I/O, stream errors, kernel support for files, System calls, library functions, file descriptors, low level file access - usage of open, creat, read, write, close, lseek, stat family, umask, dup, dup2, fcntl, file and record locking. file and directory management - chmod, chown, links(soft links & hard links - unlink, link, symlink), mkdir, rmdir, chdir, getcwd, Scanning Directories-opendir, readdir, closedir,rewinddir, seekdir, telldir functions.

UNIT III

Linux Process – Process concept, Kernel support for process, process attributes, process hierarchy,process states,process composition, process control - process creation, waiting for a process, process termination, zombie process,orphan process, system call interface for process management-fork, vfork, exit, wait, waitpid, exec family, system.Signals

Interprocess Communication : Introduction to IPC, IPC between processes on a single computer system, IPC between processes on different systems,pipes, FIFOs, Introduction to three types of IPC(Linux)-message queues, semaphores and shared memory. Message Queues.

UNI-IV

Semaphores-Kernel support for semaphores, Linux APIs for semaphores, file locking with semaphores.

Shared Memory- Kernel support for shared memory, Linux APIs for shared memory, semaphore and shared memory example.

UNIT V

Multithreaded Programming – Differences between threads and processes,Thread structure and uses,Threads and Lightweight Processes,POSIX Thread APIs, Creating Threads,Thread Attributes,Thread Synchronization with semaphores and with Mutexes,Example programs.

Sockets: Introduction to Linux Sockets, Socket system calls for connection oriented protocol and connectionless protocol, example-client/server programs.

REFERENCES:

1. Unix System Programming using C++, T.Chan, PHI.
2. Unix Concepts and Applications, 4th Edition, Sumitabha Das, TMH,2006.
3. Beginning Linux Programming, 4th Edition, N.Matthew, R.Stones,Wrox, Wiley IndiaEdition,rp-2008.
4. Linux System Programming, Robert Love, O'Reilly, SPD, rp-2007.
5. Unix Network Programming ,W.R.Stevens,PHI.
6. Unix for programmers and users, 3rd Edition, Graham Glass, King Ables, PearsonEducation, 2003.
7. Advanced Programming in the Unix environment, 2nd Edition, W.R.Stevens, PearsonEducation.
8. System Programming with C and Unix,A.Hoover,Pearson.
9. Unix Programming, Kumar Saurabh, 1st Edition, Wiley India pvt Ltd.
10. Unix and Shell programming, B.A.Forouzan and R.F.Gilberg, Cengage Learning.

Contents

UNIT : I

Linux Utilities

Introduction to Linux

Linux is a Unix-like computer operating system assembled under the model of free and open source software development and distribution. The defining component of Linux is the Linux kernel, an operating system kernel first released 5 October 1991 by Linus Torvalds.

Linux was originally developed as a free operating system for Intel x86-based personal computers. It has since been ported to more computer hardware platforms than any other operating system. It is a leading operating system on servers and other big iron systems such as mainframe computers and supercomputers more than 90% of today's 500 fastest supercomputers run some variant of Linux, including the 10 fastest. Linux also runs on embedded systems (devices where the operating system is typically built into the firmware and highly tailored to the system) such as mobile phones, tablet computers, network routers, televisions and video game consoles; the Android system in wide use on mobile devices is built on the Linux kernel.

A distribution oriented toward desktop use will typically include the X Window System and an accompanying desktop environment such as GNOME or KDE Plasma. Some such distributions may include a less resource intensive desktop such as LXDE or Xfce for use on older or less powerful computers. A distribution intended to run as a server may omit all graphical environments from the standard install and instead include other software suchas the Apache HTTP Server and an SSH server such as OpenSSH. Because Linux is freely redistributable, anyone may create a distribution for any intended use. Applications commonly used with desktop Linux systems include the Mozilla Firefox web browser, the LibreOffice office application suite, and the GIMP image editor.Since the main supporting user space system tools and libraries originated in the GNU Project, initiated in 1983 by Richard Stallman, the Free Software Foundation prefers the name *GNU/Linux*.

History of Unix

The Unix operating system was conceived and implemented in 1969 at AT&T's Bell Laboratories in the United States by Ken Thompson, Dennis Ritchie, Douglas McIlroy, and Joe Ossanna. It was first released in 1971 and was initially entirely written in assembly language, a common practice at the time. Later, in a key pioneering approach in 1973, Unix was re-written in the programming language C by Dennis Ritchie (with exceptions to the kernel and I/O). The availability of an operating system written in a high-level language allowed easier portability to different computer platforms.

Today, Linux systems are used in every domain, from embedded systems to supercomputers, and have secured a place in server installations often using the popular LAMP application stack. Use of Linux distributions in home and enterprise desktops has been growing. They have also gained popularity with various local and national governments. The federal government of Brazil is well known for its support for Linux. News of the Russian military creating its own Linux distribution has also surfaced, and has come to fruition as the G.H.ost Project. The Indian state of Kerala has gone to the extent of mandating that all state high schools run Linux on their computers.

Design

A Linux-based system is a modular Unix-like operating system. It derives much of its basic design from principles established in Unix during the 1970s and 1980s. Such a system uses a monolithic kernel, the Linux kernel, which handles process control, networking, and peripheral and file system access. Device drivers are either integrated directly with the kernel or added as modules loaded while the system is running.

Separate projects that interface with the kernel provide much of the system's higher- level functionality. The GNU userland is an important part of most Linux-based systems, providing the most common implementation of the C library, a popular shell, and many of thecommon Unix tools which carry out many basic operating system tasks. The graphical user interface (or GUI) used by most Linux systems is built on top of an implementation of the X Window System.

Programming on Linux

Most Linux distributions support dozens of programming languages. The original development tools used for building both Linux applications and operating system programs are found within the GNU toolchain, which includes the GNU Compiler Collection (GCC)and the GNU build system. Amongst others, GCC provides compilers for Ada, C, C++, Java, and Fortran. First released in 2003, the Low Level Virtual Machine project provides an alternative open-source compiler for many languages. Proprietary compilers for Linux includethe Intel C++ Compiler, Sun Studio, and IBM XL C/C++ Compiler. BASIC in the form of Visual Basic is supported in such forms as Gambas, FreeBASIC, and XBasic.

Most distributions also include support for PHP, Perl, Ruby, Python and other dynamic languages. While not as common, Linux also supports C# (via Mono), Vala, and Scheme. A number of Java Virtual Machines and development kits run on Linux, including the original Sun Microsystems JVM (HotSpot), and IBM's J2SE RE, as well as many open- source projects like Kaffe and JikesRVM.

Linux Advantages

1. **Low cost:** You don't need to spend time and money to obtain licenses since Linux and much of its software come

with the GNU General Public License. You can start to work immediately without worrying that your software may stop working anytime because thefree trial version expires. Additionally, there are large repositories from which you can freely download high quality software for almost any task you can think of.

2. **Stability:** Linux doesn't need to be rebooted periodically to maintain performance levels.It doesn't freeze up or slow down over time due to memory leaks and such. Continuous up-times of hundreds of days (up to a year or more) are not uncommon.

3. **Performance:** Linux provides persistent high performance on workstations and on networks. It can handle unusually large numbers of users simultaneously, and can make old computers sufficiently responsive to be useful again.

4. **Network friendliness:** Linux was developed by a group of programmers over the Internet and has therefore strong support for network functionality; client and server systems can be easily set up on any computer running Linux. It can perform tasks such as network backups faster and more reliably than alternative systems.

5. **Flexibility:** Linux can be used for high performance server applications, desktop applications, and embedded systems. You can save disk space by only installing the components needed for a particular use. You can restrict the use of specific computers by installing for example only selected office applications instead of the whole suite.

6. **Compatibility:** It runs all common Unix software packages and can process all commonfile formats.

7. **Choice:** The large number of Linux distributions gives you a choice. Each distribution is developed and supported by a different organization. You can pick the one you like best; the core functionalities are the same; most software runs on most distributions.

8. **Fast and easy installation:** Most Linux distributions come with user-friendly installation and setup programs. Popular Linux distributions come with tools that make installation of additional software very user friendly as well.

9. **Full use of hard disk:** Linux continues work well even when the hard disk is almost full.10.**Multitasking:** Linux is designed to do many things at the same time; e.g., a large printing job in the background won't slow down your other work.

11. **Security:** Linux is one of the most secure operating systems. "Walls" and flexible file access permission systems prevent access by unwanted visitors or viruses. Linux users have to option to select and safely download software, free of charge, from online repositories containing thousands of high quality packages. No purchase transactions requiring credit card numbers or other sensitive personal information are necessary.

12. **Open Source:** If you develop software that requires knowledge or modification of the operating system code, Linux's source code is at your fingertips. Most Linux applications are Open Source as well.

The difference between Linux and UNIX operating systems?

UNIX is copyrighted name only big companies are allowed to use the UNIX copyright and name, so IBM AIX and Sun Solaris and HP-UX all are UNIX operating systems. The Open Group holds the UNIX trademark in trust for the industry, and manages the UNIX trademarklicensing program.

Most UNIX systems are commercial in nature.

Linux is a UNIX Clone

But if you consider Portable Operating System Interface (POSIX) standards thenLinux can be considered as UNIX. To quote from Official Linux kernel README file:

Linux is a Unix clone written from scratch by Linus Torvalds with assistance from a loosely-knit team of hackers across the Net. It aims towards POSIX compliance.

However, "Open Group" do not approve of the construction "Unix-like", and consider itmisuse of their UNIX trademark.

Linux Is Just a Kernel

Linux is just a kernel. All Linux distributions includes GUI system + GNU utilities (such as cp, mv, ls,date, bash etc) + installation & management tools + GNU c/c++ Compilers + Editors (vi) + and various applications (such as OpenOffice, Firefox). However, most UNIX operating systems are considered as a complete operating system as everything come from a single source or vendor.

As I said earlier Linux is just a kernel and Linux distribution makes it complete usableoperating systems by adding various applications. Most UNIX operating systems comes with A-Z programs such as editor, compilers etc. For example HP-UX or Solaris comes with A-Z programs.

License and cost

Linux is Free (as in beer [freedom]). You can download it from the Internet orredistribute it under GNU licenses. You will see the best community support for Linux. Most UNIX like operating systems are not free (but this is changing fast, for example OpenSolaris UNIX). However, some Linux distributions such as Redhat / Novell provides additional Linux support, consultancy, bug fixing, and training for additional fees.

User-Friendly

Linux is considered as most user friendly UNIX like operating systems. It makes it easy to install sound card, flash players, and other desktop goodies. However, Apple OS X is most popular UNIX operating system for desktop usage.

Security Firewall Software

Linux comes with open source netfilter/iptables based firewall tool to protect your server and desktop from the crackers and hackers. UNIX operating systems comes with its own firewall product (for example Solaris UNIX comes with ipfilter based firewall) or you need to purchase a 3rd party software such as Checkpoint UNIX firewall.

Backup and Recovery Software

UNIX and Linux comes with different set of tools for backing up data to tape and other backup media. However, both of them share some common tools such as tar, dump/restore, and cpio etc.

File Systems

- Linux by default supports and use ext3 or ext4 file systems.
- UNIX comes with various file systems such as jfs, gpfs (AIX), jfs, gpfs (HP-UX), jfs,gpfs (Solaris).

System Administration Tools

1. UNIX comes with its own tools such as SAM on HP-UX.
2. Suse Linux comes with Yast
3. Redhat Linux comes with its own gui tools called redhat-config-*.

However, editing text config file and typing commands are most popular options for sysadmin work under UNIX and Linux.

System Startup Scripts

Almost every version of UNIX and Linux comes with system initialization script but they arelocated in different directories:

1. HP-UX - /sbin/init.d
2. AIX - /etc/rc.d/init.d
3. Linux - /etc/init.d

End User Perspective

The differences are not that big for the average end user. They will use the same shell(e.g. bash or ksh) and other development tools such as Perl or Eclipse development tool.

System Administrator Perspective

Again, the differences are not that big for the system administrator. However, youmay notice various differences while performing the following operations:

1. Software installation procedure
2. Hardware device names
3. Various admin commands or utilities
4. Software RAID devices and mirroring
5. Logical volume management
6. Package management
7. Patch management

UNIX Operating System Names

A few popular names:

1. HP-UX
2. IBM AIX
3. Sun Solairs
4. Mac OS X
5. **IRIX**

Linux Distribution (Operating System) Names

A few popular names:

1. Redhat Enterprise Linux
2. Fedora Linux
3. Debian Linux
4. Suse Enterprise Linux
5. Ubuntu Linux

Common Things Between Linux & UNIX

Both share many common applications such as:

1. GUI, file, and windows managers (KDE, Gnome)
2. Shells (ksh, csh, bash)
3. Various office applications such as OpenOffice.org
4. Development tools (perl, php, python, GNU c/c++ compilers)
5. Posix interface

FUNDAMENTAL DIFFERENCES BETWEEN LINUX AND WINDOWS

#1: Full access vs. no access

Having access to the source code is probably the single most significant difference between Linux and Windows. The fact that Linux belongs to the GNU Public License ensuresthat users (of all sorts) can access (and alter) the code to the very kernel that serves as the foundation of the Linux operating system. You want to peer at the Windows code? Goodluck. Unless you are a member of a very select (and elite, to many) group, you will never lay eyes on code making up the Windows operating system.

You can look at this from both sides of the fence. Some say giving the public access tothe code opens the operating system (and the software that runs on top of it) to malicious developers who will take advantage of any weakness they find. Others say that having full access to the code helps bring about faster improvements and bug fixes to keep those malicious developers from being able to bring the system down. I have, on occasion, dipped into the code of one Linux application or another, and when all was said and done, was happy with the results. Could I have done that with a closed-source Windows application? No.

#2: Licensing freedom vs. licensing restrictions

Along with access comes the difference between the licenses. I'm sure that every IT professional could go on and on about licensing of PC software. But let's just look at the key aspect of the licenses (without getting into legalese). With a Linux GPL-licensed operating system, you are free to modify that software and use and even republish or sell it (so long as you make the code available). Also, with the GPL, you can download a single copy of a Linux distribution (or application) and install it on as many machines as you like. With the Microsoft license, you can do none of the above. You are bound to the number of licenses you purchase, so if you purchase 10 licenses, you can legally install that operating system (or application) on only 10 machines.

#3: Online peer support vs. paid help-desk support

This is one issue where most companies turn their backs on Linux. But it's really not necessary. With Linux, you have the support of a huge community via forums, online search, and plenty of dedicated Web sites. And of course, if you feel the need, you can purchase support contracts from some of the bigger Linux companies (Red Hat and Novell for instance).

However, when you use the peer support inherent in Linux, you do fall prey to time. You could have an issue with something, send out e-mail to a mailing list or post on a forum, and within 10 minutes be flooded with suggestions. Or these suggestions could take hours of days to come in. It seems all up to chance sometimes. Still, generally speaking, most problems with Linux have been encountered and documented. So chances are good you'll find your solution fairly quickly.

On the other side of the coin is support for Windows. Yes, you can go the same route with Microsoft and depend upon your peers for solutions. There are just as many helpsites/lists/forums for Windows as there are for Linux. And you can purchase support from Microsoft itself. Most corporate higher-ups easily fall victim to the safety net that having a support contract brings. But most higher-ups haven't had to depend up on said support contract. Of the various people I know who have used either a Linux paid support contract ora Microsoft paid support contract, I can't say one was more pleased than the other. This of course begs the question "Why do so many say that Microsoft support is superior to Linux paid support?"

#4: Full vs. partial hardware support

One issue that is slowly becoming nonexistent is hardware support. Years ago, if you wanted to install Linux on a machine you had to make sure you hand-picked each piece of hardware or your installation would not work 100 percent. I can remember, back in 1997-ish, trying to figure out why I couldn't get Caldera Linux or Red Hat Linux to see my modem. After much looking around, I found I was the proud owner of a Winmodem. So I had to go out and purchase a US Robotics external modem because that was the one modem I *knew* would work. This is not so much the case now. You can grab a PC (or laptop) and most likely get one or more Linux distributions to install and work nearly 100 percent. But there are still some exceptions. For instance, hibernate/suspend remains a problem with many laptops, although it has come a long way.

With Windows, you know that most every piece of hardware will work with the operating system. Of course, there are times (and I have experienced this over and over) when you will wind up spending much of the day searching for the correct drivers for that piece of hardware you no longer have the install disk for. But you can go out and buy that 10-cent Ethernet card and know it'll work on your machine (so long as you have, or can find, the drivers). You also can rest assured that when you purchase that insanely powerful graphics card, you will probably be able to take full advantage of its power.

#5: Command line vs. no command line

No matter how far the Linux operating system has come and how amazing the desktop environment becomes, the command line will always be an invaluable tool for administration purposes. Nothing will ever replace my favorite text-based editor, ssh, and any given command-line tool. I can't imagine administering a Linux machine without the commandline. But for the end user – not so much. You could use a Linux machine for years and nevertouch the command line. Same with Windows. You can still use the command line with Windows, but not nearly to the extent as with Linux. And Microsoft tends to obfuscate the command prompt from users. Without going to Run and entering cmd (or command, or whichever it is these days), the user won't even know the command-line tool exists. And if a user does get the Windows command line up and running, how useful is it really?

#6: Centralized vs. noncentralized application installation

The heading for this point might have thrown you for a loop. But let's think about this for a second. With Linux you have (with nearly every distribution) a centralized location where you can search for, add, or remove software. I'm talking about package management systems, such as Synaptic. With Synaptic, you can open up one tool, search for an application (or group of applications), and install that application without having to do any Web searching (or purchasing).

Windows has nothing like this. With Windows, you must know where to find the software you want to install, download the software (or put the CD into your machine), and run setup.exe or install.exe with a simple double-click. For many years, it was thought that installing applications on Windows was far easier than on Linux. And for many years, that thought was right on target. Not so much now. Installation under Linux is simple, painless, and centralized.

#7: Flexibility vs. rigidity

I always compare Linux (especially the desktop) and Windows to a room where the floor and ceiling are either movable or not. With Linux, you have a room where the floor and ceiling can be raised or lowered, at will, as high or low as you want to make them. With Windows, that floor and ceiling are immovable. You can't go further than Microsoft has deemed it necessary to go.

Take, for instance, the desktop. Unless you are willing to pay for and install a third- party application that can alter the desktop appearance, with Windows you are stuck with what Microsoft has declared is the ideal desktop for you. With Linux, you can pretty much make your desktop look and feel exactly how you want/need. You can have as much or as little on your desktop as you want. From simple flat Fluxbox to a full-blown 3D Compiz experience, the Linux desktop is as flexible an environment as there is on a computer.

#8: Fanboys vs. corporate types

I wanted to add this because even though Linux has reached well beyond its school- project roots, Linux users tend to be soapbox-dwelling fanatics who are quick to spout off about why you should be choosing Linux over Windows. I am guilty of this on a daily basis (Itry hard to recruit new fanboys/girls), and it's a badge I wear proudly. Of course, this is seen as less than professional by some. After all, why would something worthy of a corporate environment have or need cheerleaders? Shouldn't the software sell itself? Because of the open source nature of Linux, it has to make do without the help of the marketing budgets and deep pockets of Microsoft. With that comes the need for fans to help spread the word. And word of mouth is the best friend of Linux.

Some see the fanaticism as the same college-level hoorah that keeps Linux in the basements for LUG meetings and science projects. But I beg to differ. Another company, thanks to the phenomenon of a simple music player and phone, has fallen into the samefanboy fanaticism, and yet that company's image has not been besmirched because of that fanaticism. Windows does not have these same fans. Instead, Windows has a league of paper- certified administrators who believe the hype when they hear the misrepresented market sharenumbers reassuring them they will be employable until the end of time.

#9: Automated vs. nonautomated removable media

I remember the days of old when you had to mount your floppy to use it and unmount it to remove it. Well, those times are drawing to a close — but not completely. One issue that plagues new Linux users is how removable media is used. The idea of having to manually "mount" a CD drive to access the contents of a CD is completely foreign to new users. There is a reason this is the way it is. Because Linux has always been a multiuser platform, it was thought that forcing a user to mount a media to use it would keep the user's files from being overwritten by another user. Think about it: On a multiuser system, if everyone had instant access to a disk that had been inserted, what would stop them from deleting or overwriting a file you had just added to the media? Things have now evolved to the point where Linux subsystems are set up so that you can use a removable device in the same way you use them in Windows. But it's not the norm. And besides, who doesn't want to manually edit the */etc/fstab* fle?

#10: Multilayered run levels vs. a single-layered run level

I couldn't figure out how best to title this point, so I went with a description. What I'mtalking about is Linux' inherent ability to stop at different run levels. With this, you can work from either the command line (run level 3) or the GUI (run level 5). This can really save your socks when X Windows is fubared and you need to figure out the problem. You can do this by booting into run level 3, logging in as root, and finding/fixing the problem.

With Windows, you're lucky to get to a command line via safe mode — and then you may or may not have the tools you need to fix the problem. In Linux, even in run level 3, you can stillget and install a tool to help you out (hello apt-get install APPLICATION via the command line). Having different run levels is helpful in another way. Say the machine in question is a Web or mail server. You want to give it all the memory you have, so you don't want the machine to boot into run level 5. However, there are times when you do want the GUI for administrative purposes (even though you can fully administer a Linux server from the command line). Because you can run the *startx* comm and from the command line at run level 3, you can still start up X Windows and have your GUI as well. With Windows, you are stuckat the Graphical run level unless you hit a serious problem.

FILE HANDLING UTILITIES:

cat Command:

cat linux command concatenates files and print it on the standard output.

SYNTAX:

The Syntax is

    cat [OPTIONS] [FILE]...

OPTIONS:

-A Show all.

-b Omits line numbers for blank space in the output.

-e A $ character will be printed at the end of each line prior to a new line.

-E Displays a $ (dollar sign) at the end of each line.

-n Line numbers for all the output lines.

-s If the output has multiple empty lines it replaces it with one empty line.

-T Displays the tab characters in the output.

-v Non-printing characters (with the exception of tabs, new-lines and form- feeds) are printed visibly.

EXAMPLE:

1. To Create a new file:cat > file1.txt

   This command creates a new file file1.txt. After typing into the file press control+d(^d) simultaneously to end the file.

2. To Append data into the file:cat >> file1.txt

   To append data into the same file use append operator >> to write into the file, else thefile will be overwritten (i.e., all of its contents will be erased).

3. To display a file:cat file1.txt

   This command displays the data in the file.

4. To concatenate several files and display:cat file1.txt file2.txt

   The above cat command will concatenate the two files (file1. txt and file2.txt) and it will display the output in the screen. Some times the output may not fit the monitor screen. In such situation you can print those files in a new file or display the file using less command.

   cat file1.txt file2.txt | less

5. To concatenate several files and to transfer the output to another file.cat file1.txt file2.txt > file3.txt

   In the above example the output is redirected to new file file3.txt. The cat command will create new file file3.txt and store the concatenated output into file3.txt.

rm Command:

rm linux command is used to remove/delete the file from the directory.

SYNTAX:

The Syntax is

s rm [options..] [file | directory]

OPTIONS:

-f Remove all files in a directory without prompting the user.

-i Interactive. With this option, rm prompts for confirmation before removing any files.

-r (or) -RRecursively remove directories and subdirectories in the argument list. The directory will be emptied of files and removed. The user is normally prompted for removal of any write-protected files which the directory contains.

EXAMPLE:

1. To Remove / Delete a file:rm file1.txt

   Here rm command will remove/delete the file file1.txt.

2. To delete a directory tree:rm -ir tmp

   This rm command recursively removes the contents of all subdirectories of the tmp directory, prompting you regarding the removal of each file, and then removes the tmpdirectory itself.

3. To remove more files at oncerm file1.txt file2.txt

   rm command removes file1.txt and file2.txt files at the same time.

cd Command:

cd command is used to change the directory.

SYNTAX:

The Syntax is

cd [directory | ~ | ./ | ../ | -]

OPTIONS:

-L Use the physical directory structure.

-P Forces symbolic links.

EXAMPLE:

1. cd linux-command

   This command will take you to the sub-directory(linux-command) from its parentdirectory.

2. cd ..

   This will change to the parent-directory from the current working directory/sub-directory.

3. cd ~

   This command will move to the user's home directory which is "/home/username".

cp Command:

cp command copy files from one location to another. If the destination is an existing file, then the file is overwritten; if the destination is an existing directory, the file is copied into the directory (the directory is not overwritten).

SYNTAX:

The Syntax is

cp [OPTIONS]... SOURCE DEST

cp [OPTIONS]... SOURCE... DIRECTORY

cp [OPTIONS]... --target-directory=DIRECTORY SOURCE...

OPTIONS:

-a same as -dpR.

-- backup[=CONTROL]make a backup of each existing destination file

-b like --backup but does not accept an argument.

-f if an existing destination file cannot be opened, remove it and try again.

-p same as --preserve=mode,ownership,timestamps.

-- preserve[=ATTR_LIST] preserve the specified attributes (default: mode,ownership,timestamps) and security contexts, if possible additional attributes: links, all.

--no-

preserve=ATTR_LISTdon't preserve the specified attribute.

--parents append source path to DIRECTORY.

EXAMPLE:

1. Copy two files:cp file1 file2

   The above cp command copies the content of file1.php to file2.php.

2. To backup the copied file:cp -b file1.php file2.php

   Backup of file1.php will be created with '~' symbol as file2. php~.

3. Copy folder and subfolders:cp -R scripts scripts1

   The above cp command copy the folder and subfolders from scripts to scripts1.

ls Command:

ls command lists the files and directories under current working directory.

SYNTAX:

The Syntax is

ls [OPTIONS]... [FILE]

OPTIONS:

-l Lists all the files, directories and their mode, Number of links, owner of the file, file size, Modified date and time and filename.

-t Lists in order of last modification time.

-a Lists all entries including hidden files.

-d Lists directory files instead of contents.

-p Puts slash at the end of each directories.

-u List in order of last access time.

-i Display inode information.

-ltr List files order by date.

-lSr List files order by file size.

EXAMPLE:

1. Display root directory contents:

   ls /

   lists the contents of root directory.

2. Display hidden files and directories:

   ls -a

   lists all entries including hidden files and directories.

3. Display inode information:

   ls -i

   7373073 book.gif

   7373074 clock.gif

   7373082 globe.gif

   7373078 pencil.gif

   7373080 child.gif

   7373081 email.gif

   7373076 indigo.gif

   The above command displays filename with inode value.

ln Command:

ln command is used to create link to a file (or) directory. It helps to provide soft link fordesired files. Inode will be different for source and destination.

SYNTAX:

The Syntax is

ln [options] existingfile(or directory)name newfile(or directory) name

OPTIONS:

-f Link files without questioning the user, even if the mode of target forbids writing. This is the default if the standard input is not a terminal.

-n Does not overwrite existing files.

-s Used to create soft links.

EXAMPLE:

1. ln -s file1.txt file2.txt

   Creates a symbolic link to 'file1.txt' with the name of 'file2.txt'. Here inode for'file1.txt' and 'file2.txt' will be different.

2. ln -s nimi nimi1

   Creates a symbolic link to 'nimi' with the name of 'nimi1'.

mkdir COMMAND:

mkdir command is used to create one or more directories.

SYNTAX:

The Syntax is

mkdir [options] directories

OPTIONS:

-m Set the access mode for the new directories.

-p Create intervening parent directories if they don't exist.

-v Print help message for each directory created.

EXAMPLE:

1. Create directory:mkdir test

   The above command is used to create the directory 'test'.

2. Create directory and set permissions:

   mkdir -m 666 test

   The above command is used to create the directory 'test' and set the read and writepermission.

rmdir Command:

rmdir command is used to delete/remove a directory and its subdirectories.

SYNTAX:

The Syntax is

rmdir [options..] Directory

OPTIONS:

-p Allow users to remove the directory dirname and its parent directories which become empty.

EXAMPLE:

1. To delete/remove a directoryrmdir tmp

   rmdir command will remove/delete the directory tmp if the directory is empty.

2. To delete a directory tree:rm -ir tmp

   This command recursively removes the contents of all subdirectories of the tmpdirectory, prompting you regarding the removal of each file, and then removes the tmpdirectory itself.

mv Command:

mv command which is short for move. It is used to move/rename file from one directory toanother. mv command is different from cp command as it completely removes the file from the source and moves to the directory specified, where cp command just copies the content from one file to another.

SYNTAX:

The Syntax is

mv [-f] [-i] oldname newname

OPTIONS:

-f This will not prompt before overwriting (equivalent to --reply=yes). mv -f will move the file(s) without prompting even if it is writing over an existing target.

-i Prompts before overwriting another file.

EXAMPLE:

1. To Rename / Move a file:

   mv file1.txt file2.txt

   This command renames file1.txt as file2.txt

2. To move a directorymv hscripts tmp

   In the above line mv command moves all the files, directories and sub-directories from hscripts folder/directory to tmp directory if the tmp directory already exists. If there is no tmp directory it rename's the hscripts directory as tmp directory.

3. To Move multiple files/More files into another directorymv file1.txt tmp/file2.txt newdir

   This command moves the files file1.txt from the current directory and file2.txt from the tmp folder/directory to newdir.

diff Command:

diff command is used to find differences between two files.

SYNTAX:

The Syntax is

diff [options..] from-file to-file

OPTIONS:

-a Treat all files as text and compare them line-by-line.

-b Ignore changes in amount of white space.

-c Use the context output format.

-e Make output that is a valid ed script.

-H Use heuristics to speed handling of large files that have numerous scattered small changes.

-i Ignore changes in case; consider upper- and lower-case letters equivalent.

-n Prints in RCS-format, like -f except that each command specifies the number of lines affected.

-q Output RCS-format diffs; like -f except that each command specifies the number of lines affected.

-r When comparing directories, recursively compare any subdirectories found.

-s Report when two files are the same.

-w Ignore white space when comparing lines.

-y Use the side by side output format.

EXAMPLE:

Lets create two files file1.txt and file2.txt and let it have the following data.

Data in file1.txt	Data in file2.txt
HIOX TEST	HIOX TEST
hscripts.com	HSCRIPTS.com
with friend ship	with friend ship

hiox india

1. Compare files ignoring white space:diff -w file1.txt file2.txt

   This command will compare the file file1.txt with file2.txt ignoring white/blank spaceand it will produce the following output.

   2c2

   < hscripts.com

   >HSCRIPTS.com

   4d3

   < Hioxindia.com

2. Compare the files side by side, ignoring white space:diff -by file1.txt file2.txt

   This command will compare the files ignoring white/blank space, It is easier todifferentiate the files.

HIOX TEST	HIOX TEST
hscripts.com	| HSCRIPTS.com
with friend ship	with friend ship
Hioxindia.com	<

   The third line(with friend ship) in file2.txt has more blank spaces, but still the -b ignores the blank space and does not show changes in the particular line, -y printout the result side by side.

3. Compare the files ignoring case.diff -iy file1.txt file2.txt

   This command will compare the files ignoring case(upper-case and lower-case) and displays the following output.

HIOX TEST	HIOX TEST
hscripts.com	HSCRIPTS.com

About wc

Short for word count, wc displays a count of lines, words, and characters in a file.

Syntax

wc [-c | -m | -C] [-l] [-w] [file ...]

-c Count bytes.

-m Count characters.

-C Same as -m.

-l Count lines.

-w Count words delimited by white space characters or new line characters. Delimiting characters are Extended Unix Code (EUC) characters from any code set defined by iswspace()

FileName of file to word count.

Examples

wc myfile.txt - Displays information about the file myfile.txt. Below is an example of theoutput.

5 13 57 myfile.txt

5 = Lines

13 = Words

57 = Characters

About split

Split a file into pieces.

Syntax

split [-linecount | -l linecount] [-a suffixlength] [file [name]]split -b n [k | m] [-a suffixlength] [file [name]]

-linecount | -llinecount Number of lines in each piece. Defaults to 1000 lines.

-a suffixlengthUse suffixlength letters to form the suffix portion of the filenames of the split file. If -a is not specified, the default suffix length is 2. If the sum of the nameoperand and the suffixlength option-argument would create a filename exceeding NAME_MAX bytes, an error will result; split will exit with a diagnostic message and no files will be created.

-b n Split a file into pieces n bytes in size.

-b n k Split a file into pieces n*1024 bytes in size.

-b n m Split a file into pieces n*1048576 bytes in size.

File The path name of the ordinary file to be split. If no input file is given or file is -, the standard input will be used.

name The prefix to be used for each of the files resulting from the split operation. Ifno name argument is given, x will be used as the prefix of the output files. The combined length of the basename of prefix and suffixlength cannot exceed NAME_MAX bytes; see OPTIONS.

Examples

split -b 22 newfile.txt new - would split the file "newfile.txt" into three separate files called newaa, newab and newac each file the size of 22.

split -l 300 file.txt new - would split the file "newfile.txt" into files beginning with the name"new" each containing 300 lines of text each

About settime and touch

Change file access and modification time.

Syntax

touch [-a] [-c] [-m] [-r ref_file | -t time] file

settime [-f ref_file] file

-a Change the access time of file. Do not change the modification time unless -m is also specified.

-c	Do not create a specified file if it does not exist. Do not write any diagnosticmessages concerning this condition.
-m	Change the modification time of file. Do not change the access time unless -ais also specified.
-r ref_file	Use the corresponding times of the file named by ref_file instead of thecurrent time.
-t time	Use the specified time instead of the current time. time will be a decimal number of the form: [[CC]YY]MMDDhhmm [.SS] MM - The month of the year [01-12]. DD - The day of the month [01-31]. hh - The hour of the day [00-23].mm - The minute of the hour [00-59]. CC - The first two digits of the year. YY - The second two digits of the year.SS - The second of the minute [00-61].
-f ref_file	Use the corresponding times of the file named by ref_file instead of thecurrent time.
File	A path name of a file whose times are to be modified.

Examples

settime myfile.txt

Sets the file myfile.txt as the current time / date.

touch newfile.txt

Creates a file known as "newfile.txt", if the file does not already exist. If the file already existsthe accessed / modification time is updated for the file newfile.txt

About comm

Select or reject lines common to two files.

Syntax

comm [-1] [-2] [-3] file1 file2

-1 Suppress the output column of lines unique to file1.

-2 Suppress the output column of lines unique to file2.

-3 Suppress the output column of lines duplicated in file1 and file2.

file1 Name of the first file to compare.

file2 Name of the second file to compare.

Examples

comm myfile1.txt myfile2.txt

The above example would compare the two files myfile1.txt and myfile2.txt.

SECURITY BY FILE PERMISSIONS

This lesson will cover the following commands:

- chmod - modify file access rights
- su - temporarily become the superuser
- chown - change file ownership
- chgrp - change a file's group ownership

File permissions

Linux uses the same permissions scheme as Linux. Each file and directory on your system is assigned access rights for the owner of the file, the members of a group of related users, and everybody else. Rights can be assigned to read a file, to write a file, and to execute a file (i.e.,run the file as a program).

To see the permission settings for a file, we can use the lscommand as follows:

```
[me@linuxbox me]$ ls -l some_file
```

-rw-rw-r-- 1 me me 1097374 Sep 26 18:48 some_file

We can determine a lot from examining the results of this command:

- The file "some_file" is owned by user "me"
- User "me" has the right to read and write this file
- The file is owned by the group "me"
- Members of the group "me" can also read and write this file
- Everybody else can read this file

Let's try another example. We will look at thebashprogram which is located in the /bin

directory:

```
[me@linuxbox me]$ ls -l /bin/bash
```

-rwxr-xr-x 1 root root 316848 Feb 27 2000 /bin/bashHere we can see:

- The file "/bin/bash" is owned by user "root"
- The superuser has the right to read, write, and execute this file
- The file is owned by the group "root"
- Members of the group "root" can also read and execute this file
- Everybody else can read and execute this file

In the diagram below, we see how the first portion of the listing is interpreted. It consists of acharacter indicating the file type, followed by three sets of three characters that convey the reading, writing and execution permission for the owner, group, and everybody else.

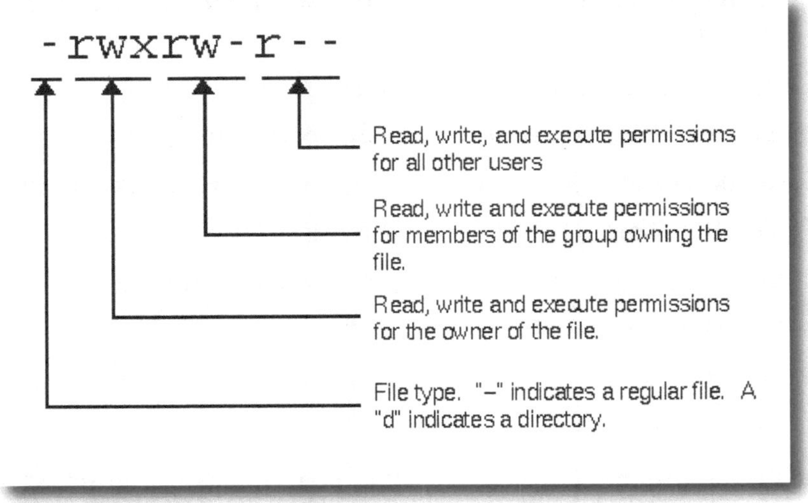

chmod

The chmod command is used to change the permissions of a file or directory. To use it, you specify the desired permission settings and the file or files that you wish to modify. There aretwo ways to specify the permissions, but I am only going to teach one way.

It is easy to think of the permission settings as a series of bits (which is how the computerthinks about them). Here's how it works:

rwx rwx rwx = 111 111 111

rw- rw- rw- = 110 110 110

rwx= 111 000 000

and so on...

rwx = 111 in binary = 7

rw- = 110 in binary = 6

r-x = 101 in binary = 5

r--= 100 in binary = 4

Now, if you represent each of the three sets of permissions (owner, group, and other) as a single digit, you have a pretty convenient way of expressing the possible permissions settings.For example, if we wanted to set some_fileto have read and write permission for the owner, but wanted to keep the file private from others, we would:

[me@linuxbox me]$chmod 600 some_file

Here is a table of numbers that covers all the common settings. The ones beginning with "7"are used with programs (since they enable execution) and the rest are for other kinds of files.

Value	*Meaning*
777	*(rwxrwxrwx)* No restrictions on permissions. Anybody may do anything. Generallynot a desirable setting.
755	*(rwxr-xr-x)* The file's owner may read, write, and execute the file. All others may read and execute the file. This setting is common for programs that are used by allusers.
700	*(rwx --------)* The file's owner may read, write, and execute the file. Nobody else has any rights. This setting is useful for programs that only the owner may use and mustbe kept private from others.
666	*(rw-rw-rw-)* All users may read and write the file.
644	*(rw-r--r--)* The owner may read and write a file, while all others may only read thefile. A common setting for data files that everybody may read, but only the owner may change.
600	*(rw ----------)* The owner may read and write a file. All others have no rights. A common setting for data files that the owner wants to keep private.

Directory permissions

The chmod command can also be used to control the access permissions for directories. In most ways, the permissions scheme for directories works the same way as they do with files. However, the execution permission is used in a different way. It provides control for access tofile listing and other things. Here are some useful settings for directories:

Value	*Meaning*
777	*(rwxrwxrwx)* No restrictions on permissions. Anybody may list files, create new filesin the directory and delete files in the directory. Generally not a good setting.
755	*(rwxr-xr-x)* The directory owner has full access. All others may list the directory, butcannot create files nor delete them. This setting is common for directories that you wish to share with other users.
700	*(rwx --------)* The directory owner has full access. Nobody else has any rights. This setting is useful for directories that only the owner may use and must be kept privatefrom others.

Becoming the superuser for a short while

It is often useful to become the superuser to perform important system administration tasks, but as you have been warned (and not just by me!), you should not stay logged on as the superuser. In most distributions, there is a program that can give you temporary access to thesuperuser's privileges. This program is called su(short for substitute user) and can be used inthose cases when you need to be the superuser for a small number of tasks. To become the superuser, simply type the sucommand. You will be prompted for the superuser's password:

```
[me@linuxbox me]$ su
Password:
[root@linuxbox me]#
```

After executing the sucommand, you have a new shell session as the superuser. To exit thesuperuser session, type exitand you will return to your previous session.

In some distributions, most notably Ubuntu, an alternate method is used. Rather than using su, these systems employ the sudo command instead. With sudo, one or more users are granted superuser privileges on an as needed basis. To execute a command as the superuser,the desired command is simply preceeded with thesudo command. After the command is entered, the user is prompted for the user's password rather than the superuser's:

```
[me@linuxbox me]$ sudo some_command
Password:
[me@linuxbox me]$
```

Changing file ownership

You can change the owner of a file by using thechowncommand. Here's an example: Suppose I wanted to change the owner ofsome_filefrom "me" to "you". I could:

```
[me@linuxbox me]$ su
Password:
[root@linuxbox me]# chown you some_file
[root@linuxbox me]# exit
[me@linuxbox me]$
```

Notice that in order to change the owner of a file, you must be the superuser. To do this, ourexample employed the sucommand, then we executed chown, and finally we typed exit to return to our previous session.

chownworks the same way on directories as it does on files.

Changing group ownership

The group ownership of a file or directory may be changed with chgrp. This command isused like this:

```
[me@linuxbox me]$ chgrp new_group some_file
```

In the example above, we changed the group ownership of some_file from its previous group to "new_group". You must be the owner of the file or directory to perform a chgrp.

chown Command:

chown command is used to change the owner / user of the file or directory. This is anadmin command, root user only can change the owner of a file or directory.

SYNTAX:

The Syntax is

chown [options] newowner filename/directoryname

OPTIONS:

- -R Change the permission on files that are in the subdirectories of the directory that you are currently in.
- -c Change the permission for each file.
- -f Prevents chown from displaying error messages when it is unable to changethe ownership of a file.

EXAMPLE:

1. chown hiox test.txt

   The owner of the 'test.txt' file is root, Change to new user hiox.

2. chown -R hiox test

   The owner of the 'test' directory is root, With -R option the files and subdirectoriesuser also gets changed.

3. chown -c hiox calc.txt

   Here Change The Owner For The Specific 'Calc.Txt' File Only.

chmod Command:

chmod command allows you to alter / Change access rights to files and directories.

File Permission is given for users,group and others as,

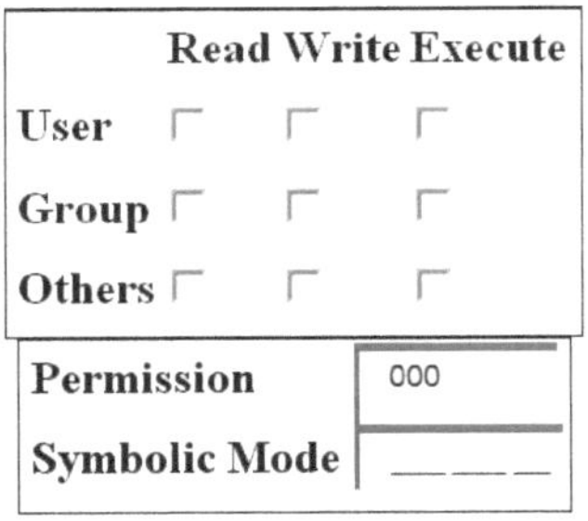

SYNTAX:

The Syntax is

chmod [options] [MODE] FileName

File Permission

#	File Permission
0	none
1	execute only
2	write only
3	write and execute
4	read only
5	read and execute
6	read and write
7	set all permissions

OPTIONS:

-c Displays names of only those files whose permissions are being changed

-f Suppress most error messages

-R Change files and directories recursively

-v Output version information and exit.

EXAMPLE:

1. To view your files with what permission they are:ls -alt

   This command is used to view your files with what permission they are.

2. To make a file readable and writable by the group and others. chmod 066 file1.txt

3. To allow everyone to read, write, and execute the filechmod 777 file1.txt

   with friend ship | with friend ship

chgrp Command:

chgrp command is used to change the group of the file or directory. This is an admincommand. Root user only can change the group of the file or directory.

SYNTAX:

The Syntax is

chgrp [options] newgroup filename/directoryname

OPTIONS:

-R Change the permission on files that are in the subdirectories of the directory that you are currently in.

-c Change the permission for each file.

-f Force. Do not report errors.

Hioxindia.com<

EXAMPLE:

1. chgrp hiox test.txt

   The group of 'test.txt' file is root, Change to newgroup hiox.

2. chgrp -R hiox test

   The group of 'test' directory is root. With -R, the files and its subdirectories alsochanges to newgroup hiox.

3. chgrp -c hiox calc.txt

   They above command is used to change the group for the specific file('calc.txt') only.

PROCESS UTILITIES:

ps COMMAND:

ps command is used to report the process status. ps is the short name for Process Status.

SYNTAX:

The Syntax isps [options]

OPTIONS:

-a	List information about all processes most frequently requested: all those except process group leaders and processes not associated with a terminal..
-A or e	List information for all processes.
-d	List information about all processes except session leaders.
-e	List information about every process now running.
-f	Generates a full listing.
-j	Print session ID and process group ID.
-l	Generate a long listing.

EXAMPLE:

1. ps

   Output:

   PID TTY   TIME CMD

   2540 pts/1   00:00:00 bash

   2621 pts/1   00:00:00 ps

   In the above example, typing ps alone would list the current running processes.

2. ps -f

   Output:

   UID PID PPID C STIME TTY   TIME CMD

   nirmala 2540 2536 0 15:31 pts/1   00:00:00 bash

   nirmala 2639 2540 0 15:51 pts/100:00:00 ps -f Displays full information about currently running processes.

kill COMMAND:

kill command is used to kill the background process.

SYNTAX:

The Syntax is

kill [-s] [-l] %pid

OPTIONS:

-s Specify the signal to send. The signal may be given as a signal name or number.

-l Write all values of signal supported by the implementation, if no operand isgiven.

-pid Process id or job id.

-9 Force to kill a process.

EXAMPLE:

Step by Step process:

- Open a process music player.xmms

  press ctrl+z to stop the process.
- To know group id or job id of the background task.jobs -l
- It will list the background jobs with its job id as,
- xmms 3956kmail 3467
- To kill a job or process.kill 3956

  kill command kills or terminates the background process xmms.

About nice

Invokes a command with an altered scheduling priority.

Syntax

nice [-increment | -n increment] command [argument ...]

-increment | -n increment	increment must be in the range 1-19; if not specified, an increment of 10 is assumed. An increment greater than 19 is equivalent to 19. The super-user may run commands with priority higher than normal by using a negative increment such as -10. A negative increment assigned by anunprivileged user is ignored.
command	The name of a command that is to be invoked. If command names any of the special built-in utilities, the results are undefined.
argument	Any string to be supplied as an argument when invoking command.

Examples

nice +13 pico myfile.txt - runs the pico command on myfile.txt with an increment of +13.About at

Schedules a command to be ran at a particular time, such as a print job late at night.

Syntax

at	executes commands at a specified time.
atq	lists the user's pending jobs, unless the user is the superuser; in that case, everybody's jobs are listed. The format of the output lines (one for each job) is: Job number, date, hour, job class.
atrm	deletes jobs, identified by their job number.
batch	executes commands when system load levels permit; in other words, when the loadaverage drops below 1.5, or the value specified in the invocation of atrun.

at [-c | -k | -s] [-f filename] [-q queuename] [-m] -t time [date] [-l] [-r]

-c	C shell. csh(1) is used to execute the at-job.

-k	Korn shell. ksh(1) is used to execute the at-job.
-s	Bourne shell. sh(1) is used to execute the at-job.
-f filename	Specifies the file that contains the command to run.
-m	Sends mail once the command has been run.
-t time	Specifies at what time you want the command to be ran. Format hh:mm. am / pm indication can also follow the time otherwise a 24-hour clock is used. A timezone name of GMT, UCT or ZULU (case insensitive) can follow to specify that the time is in Coordinated Universal Time. Other timezones can be specified using the TZ environment variable. The below quick times can also be entered: midnight - Indicates the time 12:00 am (00:00). noon - Indicates the time 12:00 pm. now - Indicates the current day and time. Invoking at - now will submitsubmit an at-job for potentially immediate execution.
date	Specifies the date you wish it to be ran on. Format month, date, year. Thefollowing quick days can also be entered: today - Indicates the current day. tomorrow - Indicates the day following the current day.
-l	Lists the commands that have been set to run.
-r	Cancels the command that you have set in the past.

Examples

at -m 01:35 < atjob = Run the commands listed in the 'atjob' file at 1:35AM, in addition all output that is generated from job mail to the user running the task. When this command has been successfully enter you should receive a prompt similar to the below example.

commands will be executed using /bin/cshjob 1072250520.a at Wed Dec 24 00:22:00 2003

at -l = This command will list each of the scheduled jobs as seen below.1072250520.a Wed Dec 24 00:22:00 2003

at -r 1072250520.a = Deletes the job just created.or

atrm 23 = Deletes job 23.

If you wish to create a job that is repeated you could modify the file that executes the commands with another command that recreates the job or better yet use the crontab command.

FILTERS:

more COMMAND:

more command is used to display text in the terminal screen. It allows only backwardmovement.

SYNTAX:

The Syntax is

more [options] filename

OPTIONS:

-c	Clear screen before displaying.
-e	Exit immediately after writing the last line of the last file in the argument list.
-n	Specify how many lines are printed in the screen for a given file.
+n	Starts up the file from the given number.

EXAMPLE:

1. more -c index.php

   Clears the screen before printing the file .

2. more -3 index.php

Prints first three lines of the given file. Press **Enter** to display the file line by line.

head COMMAND:

head command is used to display the first ten lines of a file, and also specifies how manylines to display.

SYNTAX:

The Syntax is

head [options] filename

OPTIONS:

-n	To specify how many lines you want to display.
-n number	The number option-argument must be a decimal integer whose sign affects the location in the file, measured in lines.
-c number	The number option-argument must be a decimal integer whose sign affects the location in the file, measured in bytes.

EXAMPLE:

1. head index.php

   This command prints the first 10 lines of 'index.php'.

2. head -5 index.php

   The head command displays the first 5 lines of 'index.php'.

3. head -c 5 index.php

   The above command displays the first 5 characters of 'index. php'.

tail COMMAND:

tail command is used to display the last or bottom part of the file. By default it displayslast 10 lines of a file.

SYNTAX:

The Syntax is

tail [options] filename

OPTIONS:

-l	To specify the units of lines.
-b	To specify the units of blocks.
-n	To specify how many lines you want to display.
-c number	The number option-argument must be a decimal integer whose sign affects the location in the file, measured in bytes.
-n number	The number option-argument must be a decimal integer whose sign affects the location in the file, measured in lines.

EXAMPLE:

1. tail index.php

   It displays the last 10 lines of 'index.php'.

2. tail -2 index.php

   It displays the last 2 lines of 'index.php'.

3. tail -n 5 index.php

   It displays the last 5 lines of 'index.php'.

4. tail -c 5 index.php

   It displays the last 5 characters of 'index.php'.

cut COMMAND:

cut command is used to cut out selected fields of each line of a file. The cut command uses delimiters to determine where to split fields.

SYNTAX:

The Syntax is cut [options]

OPTIONS:

-c	Specifies character positions.
-b	Specifies byte positions.
-d flags	Specifies the delimiters and fields.

EXAMPLE:

1. cut -c1-3 text.txt

   Output:

   Thi

   Cut the first three letters from the above line.

2. cut -d, -f1,2 text.txt

   Output:

   This is, an example program

   The above command is used to split the fields using delimiter and cut the first twofields.

paste COMMAND:

paste command is used to paste the content from one file to another file. It is also used toset column format for each line.

SYNTAX:

The Syntax is paste [options]

OPTIONS:

-s	Paste one file at a time instead of in parallel.
-d	Reuse characters from LIST instead of TABs .

EXAMPLE:

1. paste test.txt>test1.txt

   Paste the content from 'test.txt' file to 'test1.txt' file.

2. ls | paste - - - -

List all files and directories in four columns for each line.

sort COMMAND:

sort command is used to sort the lines in a text file.

SYNTAX:

The Syntax is

sort [options] filename

OPTIONS:

-r	Sorts in reverse order.
-u	If line is duplicated display only once.
-o filename	Sends sorted output to a file.

EXAMPLE:

1. sort test.txt
   Sorts the 'test.txt'file and prints result in the screen.
2. sort -r test.txt
   Sorts the 'test.txt' file in reverse order and prints result in the screen.

About uniq

Report or filter out repeated lines in a file.

Syntax

uniq [-c | -d | -u] [-f fields] [-s char] [-n] [+m] [input_file [output_file]]

-c	Precede each output line with a count of the number of times the lineoccurred in the input.
-d	Suppress the writing of lines that are not repeated in the input.
-u	Suppress the writing of lines that are repeated in the input.

-f fields	Ignore the first fields fields on each input line when doing comparisons,where fields is a positive decimal integer. A field is the maximal string matched by the basic regular expression: [[:blank:]]*[^[:blank:]]* If fields specifies more fields than appear on an input line, a null string willbe used for comparison.
-s char	Ignore the first chars characters when doing comparisons, where chars is apositive decimal integer. If specified in conjunction with the -f option, the first chars characters after the first fields fields will be ignored. If chars specifies more characters than remain on an input line, a null string will be used for comparison.
-n	Equivalent to -f fields with fields set to n.
+m	Equivalent to -s chars with chars set to m.
input_file	A path name of the input file. If input_file is not specified, or if the input_file is -, thestandard input will be used.
output_file	A path name of the output file. If output_file is not specified, the standard output will be used. The results are unspecified if the file named by output_file is the file named by input_file.

Examples

uniq myfile1.txt > myfile2.txt - Removes duplicate lines in the first file1.txt and outputs theresults to the second file.

About tr

Translate characters.

Syntax

tr [-c] [-d] [-s] [string1] [string2]

-c	Complement the set of characters specified by string1.
-d	Delete all occurrences of input characters that are specified by string1.
-s	Replace instances of repeated characters with a single character.
string1	First string or character to be changed.
string2	Second string or character to change the string1.

Examples

echo "12345678 9247" | tr 123456789 computerh - this example takes an echo response of '12345678 9247' and pipes it through the tr replacing the appropriate numbers with the letters.In this example it would return *computer hope.*

tr -cd '\11\12\40-\176' < myfile1 > myfile2 - this example would take the file myfile1 and strip all non printable characters and take that results to myfile2.

General Commands:

date COMMAND:

date command prints the date and time.

SYNTAX:

The Syntax is

date [options] [+format] [date]

OPTIONS:

-a	Slowly adjust the time by sss.fff seconds (fff represents fractions of a second). This adjustment can be positive or negative.Only system admin/ super user can adjust the time.
- s date-string	Sets the time and date to the value specfied in the datestring. The datestr may contain the month names, timezones, 'am', 'pm', etc.

-u Display (or set) the date in Greenwich Mean Time (GMT-universal time).

Format:

%a Abbreviated weekday(Tue).

%A Full weekday(Tuesday).

%b Abbreviated month name(Jan).

%B Full month name(January).

%c Country-specific date and time format..

%D Date in the format %m/%d/%y.

%j Julian day of year (001-366).

%n Insert a new line.

%p String to indicate a.m. or p.m.

%T Time in the format %H:%M:%S.

%t Tab space.

%V Week number in year (01-52); start week on Monday.

EXAMPLE:

1. date commanddate

   The above command will print Wed Jul 23 10:52:34 IST 2008

2. To use tab space:

   date +"Date is %D %t Time is %T"

   The abovecommand will remove space and print asDate is 07/23/08 Time is 10:52:34

3. To know the week number of the year,date -V

   The above command will print 30

4. To set the date,

date -s "10/08/2008 11:37:23"

The above command will print Wed Oct 08 11:37:23 IST 2008

who COMMAND:

who command can list the names of users currently logged in, their terminal, the time theyhave been logged in, and the name of the host from which they have logged in.

SYNTAX:

The Syntax is

who [options] [file]

OPTIONS:

am i	Print the username of the invoking user, The 'am' and 'i' must be space separated.
-b	Prints time of last system boot.
-d	print dead processes.
-H	Print column headings above the output.
-i	Include idle time as HOURS:MINUTES. An idle time of . indicates activitywithin the last minute.
-m	Same as who am i.
-q	Prints only the usernames and the user count/total no of users logged in.
-T,-w	Include user's message status in the output.

EXAMPLE:

1. who -uH

   Output:

   NAME LINE TIME IDLE PID COMMENT

   hiox ttyp3 Jul 10 11:08 . 4578

   This sample output was produced at 11 a.m. The "." indiacates activity within the lastminute.

2. who am i

   who am i command prints the user name.

echo COMMAND:

echo command prints the given input string to standard output.

SYNTAX:

The Syntax is

    echo [options..] [string]

OPTIONS:

-n	do not output the trailing newline
-e	enable interpretation of the backslash-escaped characters listed below
-E	disable interpretation of those sequences in STRINGs

Without -E, the following sequences are recognized and interpolated:

\NNN	the character whose ASCII code is NNN (octal)
\a	alert (BEL)
\\	backslash
\b	backspace
\c	suppress trailing newline
\f	form feed
\n	new line
\r	carriage return
\t	horizontal tab
\v	vertical tab

EXAMPLE:

1. echo command

   echo "hscripts Hiox India"

The above command will print as hscripts Hiox India

2. To use backspace:

   echo -e "hscripts \bHiox \bIndia"

   The above command will remove space and print as hscriptsHioxIndia

3. To use tab space in echo commandecho -e "hscripts\tHiox\tIndia"

   The above command will print as hscripts HioxIndia

passwd COMMAND:

passwd command is used to change your password.

SYNTAX:

The Syntax is

passwd [options]

OPTIONS:

-a	Show password attributes for all entries.
-l	Locks password entry for name.
-d	Deletes password for name. The login name will not be prompted forpassword.
-f	Force the user to change password at the next login by expiring thepassword for name.

EXAMPLE:

1. passwd

   Entering just passwd would allow you to change the password. After entering passwdyou will receive the following three prompts:

   Current Password:

   New Password:

   Confirm New Password:

Each of these prompts must be entered correctly for the password to be successfullychanged.

pwd COMMAND:

pwd - Print Working Directory. pwd command prints the full filename of the currentworking directory.

SYNTAX:

The Syntax is

pwd [options]

OPTIONS:

-P	The pathname printed will not contain symbolic links.
-L	The pathname printed may contain symbolic links.

EXAMPLE:

1. Displays the current working directory.pwd

   If you are working in home directory then, pwd command displays the currentworking directory as /home.

cal COMMAND:

cal command is used to display the calendar.

SYNTAX:

The Syntax is

cal [options] [month] [year]

OPTIONS:

-1	Displays single month as output.
-3	Displays prev/current/next month output.
-s	Displays sunday as the first day of the week.
-m	Displays Monday as the first day of the week.

-j Displays Julian dates (days one-based, numbered from January 1).

-y Displays a calendar for the current year.

EXAMPLE:

1. cal

   Output:

   September 2008

   Su Mo Tu We Th Fr Sa1 2 3 4 5 6

   7 8 9 10 11 12 13

   14 15 16 17 18 19 20

   21 22 23 24 25 26 27

   28 29 30

   cal command displays the current month calendar.

2. cal -3 5 2008

   Output:

   April 2008 May 2008 June 2008

   Su Mo Tu We Th Fr Sa Su Mo Tu We Th Fr Sa Su Mo Tu We Th Fr Sa1 2 3 4 5 1 2 3 1 2 3 4 5 6 7 6 7 8 9 10 11 12 4 5 6 7 8 9 10 8 9 10 11 12 13 14 13 14 15 16 17 18 19 11 12 13 14 15 16 17 15 16 17 18 19 20 21 20 21 22 23 24 25 26 18 19 20 21 22 23 24 22 23 24 25 26 27 28 27 28 29 3025 26 27 28 29 30 31 29 30

   Here the cal command displays the calendar of April, May and June month of year2008.

login Command

Signs into a new system.

Syntax

login [-p] [-d device] [-h hostname | terminal | -r hostname] [name [environ]]

-p	Used to pass environment variables to the login shell.
-d device	login accepts a device option, device. device is taken to be the path name of the TTY port login is to operate on. The use of the device option can be expected to improve login performance, since login will not need to call ttyname. The -d option is available only to users whose UID and effective UID are root. Any other attempt to use -d will cause login to quietly exit.
-h hostname I terminal	Used by in.telnetd to pass information about the remote host and terminaltype.
-r hostname	Used by in.rlogind to pass informationaboutthe remote host. Examples

login computerhope.com - Would attempt to login to the computerhope domain.

uname command

Print name of current system.

Syntax

uname [-a] [-i] [-m] [-n] [-p] [-r] [-s] [-v] [-X] [-S systemname]

-a	Print basic information currently available from the system.
-i	Print the name of the hardware implementation (platform).
-m	Print the machine hardware name (class). Use of this option is discouraged; use uname -p instead.
-n	Print the nodename (the nodename is the name by which the system is knownto a communications network).
-p	Print the current host's ISA or processor type.
-r	Print the operating system release level.

-s	Print the name of the operating system. This is the default.
-v	Print the operating system version.
-X	Print expanded system information, one information element per line, as expected by SCO Unix. The displayed information includes: • system name, node, release, version, machine, and number of CPUs. • BusType, Serial, and Users (set to "unknown" in Solaris) • OEM# and Origin# (set to 0 and 1, respectively)
-S systemname	The nodename may be changed by specifying a system name argument. The system name argument is restricted to SYS_NMLN characters. SYS_NMLN is an implementation specific value defined in <sys/utsname.h>. Only the super-user is allowed this capability.

Examples

uname -arv

List the basic system information, OS release, and OS version as shown below. SunOS hope 5.7 Generic_106541-08 sun4m sparc SUNW,SPARCstation-10

uname -p

Display the Linux platform.

Disk utilities

df - summarize disk block and file usage

df is used to report the number of disk blocks and inodes used and free for each file system.The output format and valid options are very specific to the OS and program version in use.

Syntax

df [options] [resource]

Common Options

-l local file systems only(SVR4) **-k** report in kilobytes (SVR4) **df**

Filesystem kbytes used avail capacity Mounted on

/dev/sd0a 20895 19224 0 102% /

/dev/sd0h 319055 131293 155857 46% /usr

/dev/sd1g 637726 348809 225145 61% /usr/local

du - report disk space in use

du reports the amount of disk space in use for the files or directories you specify.

Syntax

du [options] [directory or file]

Common Options

-a display disk usage for each file, not just subdirectories **-s** display a summary total only

-k report in kilobytes (SVR4)du

1 ./.elm

1 ./Mail

1 ./News

20 ./uc du -a uc

7 uc/unixgrep.txt

5 uc/editors.txt

1 uc/.emacs

1 uc/.exrc

4 uc/telnet.ftp

1 uc/uniq.tee.txt

20 uc

NETWORKING COMMANDS

Command/Syntax	What it will do
finger [options] *user[@hostname]*	report information about users on local and remote machines
ftp [options] *host*	transfer file(s) using file transfer protocol
rcp [options] *hostname*	remotely copy files from this machine to another machine
rlogin [options] *hostname*	login remotely to another machine
rsh [options] *hostname*	remote shell to run on another machine
telnet [host [port]]	communicate with another host using telnet protocol

TELNET and **FTP** are Application Level Internet protocols. The TELNET and FTP protocolspecifications have been implemented by many different sources, including The National Center for Supercomputer Applications (NCSA), and many other public domain and shareware sources ***rlogin*** is a remote login service that was at one time exclusive to Berkeley

4.3 BSD UNIX

Essentially, it offers the same functionality as ***telnet***, except that it passes to the remote computer information about the user's login environment. Machines can be configured to allow connections from trusted hosts without prompting for the users' passwordsA more secure version of this protocol is the Secure SHell, **SSH**, software written by Tatu Ylonen and available via ftp://ftp.net. ohio-state.edu/pub/security/ssh.

their commands—rsh (remotshell), rcp (remote copy), and rlogin (remote login)—wereprevalent in the past, but because they

offer little security, they're generally discouraged in today's environments. rsh and rloginare similar in functionality to telnet, and rcp is similar to ftp.

telnet [options] [remote_host [port_number]

] ***tn3270*** [options] [remote_host [

port_number]] ***ftp*** [options] [remote_host]

Common

Options ftp

telnet Action

-d set debugging mode on

-d same as above (SVR4only) **-i** turn offinteractive prompting

-n don't attempt auto-login on connection **-v** verbose mode on

-l user connect with username, **user**, on the remote host(SVR4 only) **-8** 8-bit data path (SVR4 only)

telnet solaris or

telnet 192.168.1

telnet oscar.us.ohio-state.edu tn3270ohstmvsa.acs.ohio-state.edu ftp magnus.acs.ohio-state.edu

finger - get information about users

finger displays the **.plan** file of a specific user, or reports who is logged into a specificmachine. The user must allow general read permission on the **.plan** file.

Syntax

finger [options] [user[@hostname]]

Common Options

-l force long output format

-m match username only, not first or last names

-s force short output format

Examples

brigadier: condron [77]> finger workshop@nyssa This is a sample .plan file for the nyssa id, workshop.

This id is being used this week by Frank Fiamingo, Linda DeBula, and Linda Condron, while we teach a pilot versionof the new Unix workshop we developed for UTS.

Remote login

rlogin [-l username] remote_host

rsh [-l username] remote_host [command]

rcp [[user1]@host1:]original_filename [[user2]@host2:]new_filename

where the parts in brackets ([]) are optional. *rcp* does not prompt for passwords, so you must have permission to execute remote commands on the specified machines as the selected user on each machine.

Common Options

-l username connect as the user, **username**, on the remote host (***rlogin*** & ***rsh***)

Using ssh

ssh (Secure SHell) and telnet are two methods that enable you to log in to a remote system andrun commands interactively;

command hostname

ssh darwin or

ssh 192.168.1.58

The ping Utility

The ping command sends an echo request to a host available on the network. Using thiscommand you can check if your remote host is responding well or not.

The ping command is useful for the following –

- Tracking and isolating hardware and software problems.
- Determining the status of the network and various foreign hosts.
- Testing, measuring, and managing networks.

Syntax

Following is the simple syntax to use **ping** command –

$ping hostname or ip-address

Above command would start printing a response after every second. To come out of the command you can terminate it by pressing CNTRL + C keys.

Example

Following is the example to check the availability of a host available on the network –

$ping google.com

PING google.com (74.125.67.100) 56(84) bytes of data.

64 bytes from 74.125.67.100: icmp_seq=1 ttl=54 time=39.4 ms

64 bytes from 74.125.67.100: icmp_seq=2 ttl=54 time=39.9 ms

64 bytes from 74.125.67.100: icmp_seq=3 ttl=54 time=39.3 ms

64 bytes from 74.125.67.100: icmp_seq=4 ttl=54 time=39.1 ms

64 bytes from 74.125.67.100: icmp_seq=5 ttl=54 time=38.8 ms

--- google.com ping statistics ---

22 packets transmitted, 22 received, 0% packet loss, time 21017ms
rtt min/avg/max/mdev = 38.867/39.334/39.900/0.396 ms

$

If a host does not exist then it would behave something like this –

$ping giiiiigle.com

ping: unknown host giiiiigle.com

$

The ftp Utility

Here ftp stands for **F**ile **T**ransfer **P**rotocol. This utility helps you to upload and download yourfile from one computer to another computer.

The ftp utility has its own set of UNIX like commands which allow you to perform tasks such as–

- Connect and login to a remote host.

- Navigate directories.
- List directory contents
- Put and get files
- Transfer files as ascii, ebcdic or binary

Syntax

Following is the simple syntax to use **ping** command –

$ftp hostname or ip-address

Above command would prompt you for login ID and password. Once you are authenticated, youwould have access on the home directory of the login account and you would be able to perform various commands.

Few of the useful commands are listed below –

Command	Description
put filename	Upload filename from local machine to remote machine.
get filename	Download filename from remote machine to local machine.
mput file list	Upload more than one files from local machine to remote machine.
mget file list	Download more than one files from remote machine to local machine.
prompt off	Turns prompt off, by default you would be prompted to upload or download movies using mput or mget commands.
prompt on	Turns prompt on.
Dir	List all the files available in the current directory of remote machine.
cd dirname	Change directory to dirname on remote machine.
lcd dirname	Change directory to dirname on local machine.
Quit	Logout from the current login.

It should be noted that all the files would be downloaded or uploaded to or from current directories. If you want to upload your files in a particular directory then first you change to that directory and then upload required files.

Example

Following is the example to show few commands –

$ftp amrood.com Connected to amrood.com.

220 amrood.com FTP server (Ver 4.9 Thu Sep 2 20:35:07 CDT 2009)Name (amrood.com:amrood): amrood

331 Password required for amrood.Password:

230 User amrood logged in.ftp> dir

200 PORT command successful.

150 Opening data connection for /bin/ls.

total 1464

drwxr-sr-x	3 amrood	group	1024 Mar 11 20:04 Mail
drwxr-sr-x	2 amrood	group	1536 Mar 3 18:07 Misc
drwxr-sr-x	5 amrood	group	512 Dec 7 10:59 OldStuff
drwxr-sr-x	2 amrood	group	1024 Mar 11 15:24 bin
drwxr-sr-x	5 amrood	group	3072 Mar 13 16:10 mpl

-rw-r--r-- 1 amrood group 209671 Mar 15 10:57 myfile.out

drwxr-sr-x 3 amrood group
512 Jan 5 13:32 public

drwxr-sr-x 3 amrood group
512 Feb 10 10:17 pvm3

226 Transfer complete.ftp> cd mpl

250 CWD command successful.ftp> dir

200 PORT command successful.

150 Opening data connection for /bin/ls.total 7320

-rw-r--r-- 1 amrood group 1630 Aug 8 1994 dboard.f

-rw-r----- 1 amrood group 4340 Jul 17 1994 vttest.c

-rwxr-xr-x 1 amrood group525574 Feb 15 11:52 wave_shift

-rw-r--r-- 1 amrood group 1648 Aug 5 1994 wide.list

-rwxr-xr-x 1 amrood group4019 Feb 14 16:26 fix.c

226 Transfer complete.ftp> get wave_shift

200 PORT command successful.

150 Opening data connection for wave_shift (525574 bytes).226 Transfer complete.

528454 bytes received in 1.296 seconds (398.1 Kbytes/s)ftp> quit

221 Goodbye.

$

The telnet Utility

Many times you would be in need to connect to a remote Unix machine and work on thatmachine remotely. Telnet is a utility that allows a computer user at one site to make a connection, login and then conduct work on a computer at another site.

Once you are login using telnet, you can perform all the activities on your remotely connectmachine. Here is example telnet session –

C:>telnet amrood.comTrying...

Connected to amrood.com.

Escape character is '^]'.

login: amrood amrood's Password:

```
*****************************************************
*                                       *
*                                       *
*    WELCOME TO AMROOD.COM                        *
*                                       *
*                                       *
```

```
*************************************************

Last unsuccessful login: Fri Mar 3 12:01:09 IST 2009

Last login: Wed Mar  8 18:33:27 IST 2009 on pts/10

    { do your work }

$ logout

Connection closed.

C:>
```

The finger Utility

The finger command displays information about users on a given host. The host can be eitherlocal or remote.

Finger may be disabled on other systems for security reasons.

Following are the simple syntax to use finger command –

Check all the logged in users on local machine as follows –

```
$ finger

Login   Name    Tty          Idle Login Time Office amrood
                pts/0        Jun 25 08:03 (62.61.164.115)
```

Get information about a specific user available on local machine –

```
$ finger amrood

Login: amrood                Name: (null)

Directory: /home/amrood      Shell: /bin/bash

On since Thu Jun 25 08:03 (MST) on pts/0 from 62.61.164.115
No mail.

No Plan.
```

Check all the logged in users on remote machine as follows –

```
$ finger @avtar.com

Login   Name  Tty            Idle Login Time Office amrood
              pts/0          Jun 25 08:03 (62.61.164.115)
```

Get information about a specific user available on remote machine –

$ finger amrood@avtar.com

Login: amrood Name: (null)

Directory: /home/amrood Shell: /bin/bash

On since Thu Jun 25 08:03 (MST) on pts/0 from 62.61.164.115
No mail.

No Plan.

finding host/domain name and IP address - **hostname**

- test network connection – **ping**
- getting network configuration – **ifconfig**
- Network connections, routing tables, interface statistics – **netstat**
- query DNS lookup name – **nslookup**
- communicate with another hostname – **telnet**
- outing steps that packets take to get to network host – **traceroute**
- view user information – **finger**
- checking status of destination host - **telnet**

Computers are connected in a network to exchange information or resources each other. Two or more computer connected through network media called computer network. There are number of network devices or media are involved to form computer network. Computer loaded with Linux Operating System can also be a part of network whether it is small or large network by its multitasking and multiuser natures. Maintaining of system and network up and running is a task of System/Network Administrator's job. In this article we are going to review frequently used network configuration and troubleshoot commands in Linux.

Linux Network Configuration Commands

TEXT PROCESSING COMMANDS

sort

File sort utility, often used as a filter in a pipe. This command sorts a *text stream* or file forwards or backwards, or according to various keys or character positions. Using the -m option, it merges presorted input files. The *info page* lists its many capabilities and options.

tsort

Topological sort, reading in pairs of whitespace-separated strings and sorting accordingto input patterns. The original purpose of **tsort** was to sort a list of dependencies for an obsolete version of the *ld* linker in an «ancient» version of UNIX.

The results of a *tsort* will usually differ markedly from those of the standard **sort** command, above.

uniq

This filter removes duplicate lines from a sorted file. It is often seen in a pipe coupledwith sort.

cat list-1 list-2 list-3 | sort | uniq > final.list

Concatenates the list files,

sorts them,

removes duplicate lines,

and finally writes the result to an output file.

The useful -c option prefixes each line of the input file with its number of occurrences.

bash$ **cat testfile**

This line occurs only once.

This line occurs twice.

This line occurs twice.

This line occurs three times.

This line occurs three times.

This line occurs three times.

bash$ **uniq -c testfile**

1 This line occurs only once.

2 This line occurs twice.

3 This line occurs three times.

bash$ sort testfile | uniq -c | sort -nr

3 This line occurs three times.

2 This line occurs twice.

1 This line occurs only once.

The **sort INPUTFILE | uniq -c | sort -nr** command string produces a *frequency of occurrence* listing on the INPUTFILE file (the -nr options to **sort** cause a reversenumerical sort). This template finds use in analysis of log files and dictionary lists, and wherever the lexical structure of a document needs to be examined.

15.12. Word Frequency Analysis

&wf;

bash$ **cat testfile**

This line occurs only once.

This line occurs twice.

This line occurs twice.

This line occurs three times.

This line occurs three times.

This line occurs three times.

bash$ **./wf.sh testfile**

6 this

6 occurs

6 line

3 times

3 three

2 twice

1 only

1 once

expand, unexpand

The **expand** filter converts tabs to spaces. It is often used in a pipe.

The **unexpand** filter converts spaces to tabs. This reverses the effect of **expand**.

cut

A tool for extracting fields from files. It is similar to the **print $N** command set in awk, but more limited. It may be simpler to use *cut* in a script than *awk*. Particularly important are the -d (delimiter) and -f (field specifier) options.

Using **cut** to obtain a listing of the mounted filesystems:

cut -d ' ' -f1,2 /etc/mtab

Using **cut** to list the OS and kernel version:

uname -a | cut -d" " -f1,3,11,12

Using **cut** to extract message headers from an e-mail folder:

bash$ grep '^Subject:' read-messages | cut -c10-80

Re: Linux suitable for mission-critical apps? MAKE MILLIONS WORKING AT HOME!!!

Spam complaint

Re: Spam complaint Using **cut** to parse a file:

List all the users in /etc/passwd.

FILENAME=/etc/passwd

for user in $(cut -d: -f1 $FILENAME)do

echo $userdone

Thanks, Oleg Philon for suggesting this.

cut -d ' ' -f2,3 filename is equivalent to awk -F'[ ]' '{ print $2, $3 }' filename

Замечание

It is even possible to specify a linefeed as a delimiter. The trick is to actuallyembed a linefeed (**RETURN**) in the command sequence.

bash$ **cut -d'**

' -f3,7,19 testfile

This is line 3 of testfile.

This is line 7 of testfile.

This is line 19 of testfile.

paste

Tool for merging together different files into a single, multi-column file. In combinationwith cut, useful for creating system log files.

join

Consider this a special-purpose cousin of **paste**. This powerful utility allows merging twofiles in a meaningful fashion, which essentially creates a simple version of a relational database.

The **join** command operates on exactly two files, but pastes together only those lines witha common tagged field (usually a numerical label), and writes the result to stdout. The files to be joined should be sorted according to the tagged field for the matchups to work properly.

File: 1.data

100 Shoes

200 Laces

300 Socks

File: 2.data

100 $40.00

200 $1.00

300 $2.00

bash$ **join 1.data 2.data**

File: 1.data 2.data

100 Shoes $40.00

200 Laces $1.00

300 Socks $2.00

The tagged field appears only once in the output.

head

lists the beginning of a file to stdout. The default is 10 lines, but a different number canbe specified. The command has a number of interesting options.

Which files are scripts?

&scriptdetector;

Generating 10-digit random numbers

&rnd;

tail

lists the (tail) end of a file to stdout. The default is 10 lines, but this can be changed with the -n option. Commonly used to keep track of changes to a system logfile, using the -f option, which outputs lines appended to the file.

Using *tail* to monitor the system log

&ex12;

To list a specific line of a text file, pipe the output of **head** to **tail -n 1**. For example **head -n 8database.txt | tail -n 1** lists the 8th line of the file database.txt.

To set a variable to a given block of a text file:

var=$(head -n $m $filename | tail -n $n)# filename = name of file

m = from beginning of file, number of lines to end of block

n = number of lines to set variable to (trim from end of block)

Newer implementations of **tail** deprecate the older **tail -$LINES filename** usage. The standard

tail -n $LINES filename is correct.

A multi-purpose file search tool that uses Regular Expressions. It was originally acommand/filter in the venerable **ed** line editor: **g/re/p** -- *global - regular expression - print.*

grep *pattern* [*file...*]

Search the target file(s) for occurrences of *pattern*, where *pattern* may be literal text or a RegularExpression.

bash$ grep '[rst]ystem.$' osinfo.txt

The GPL governs the distribution of the Linux operating system.

If no target file(s) specified, **grep** works as a filter on stdout, as in a pipe.bash$ **ps ax | grep clock**

765 tty1	S	0:00 xclock
901 pts/1	S	0:00 grep clock

The -i option causes a case-insensitive search. The -w option matches only whole words.

The -l option lists only the files in which matches were found, but not the matching lines.

The -r (recursive) option searches files in the current working directory and all subdirectoriesbelow it.

The -n option lists the matching lines, together with line numbers. bash$ **grep -n Linux osinfo.txt**

2:This is a file containing information about Linux.

6:The GPL governs the distribution of the Linux operating system. The -v (or --invert-match) option *filters out* matches.

grep pattern1 *.txt | grep -v pattern2

Matches all lines in "*.txt" files containing "pattern1",

but ***not*** "pattern2".

The -c (--count) option gives a numerical count of matches, rather than actually listingthe matches.

grep -c txt *.sgml # (number of occurrences of "txt" in "*.sgml" files)

grep -cz .

^ dot

means count (-c) zero-separated (-z) items matching "."

that is, non-empty ones (containing at least 1 character).

#

printf 'a b\nc d\n\n\n\n\n\000\n\000e\000\000\nf' | grep -cz . # 3

printf 'a b\nc d\n\n\n\n\n\000\n\000e\000\000\nf' | grep -cz '$' # 5

printf 'a b\nc d\n\n\n\n\n\000\n\000e\000\000\nf' | grep -cz '^' # 5 #

printf 'a b\nc d\n\n\n\n\n\000\n\000e\000\000\nf' | grep -c '$' # 9

By default, newline chars (\n) separate items to match.

Note that the -z option is GNU "grep" specific.

Thanks, S.C.

The --color (or --colour) option marks the matching string in color (on the console or in an *xterm* window). Since *grep* prints out each entire line containing the matching pattern, this lets you see exactly *what* is being matched. See also the -o option, which shows only the matching portion of the line(s).

Printing out the *From* lines in stored e-mail messages

&fromsh;

When invoked with more than one target file given, **grep** specifies which file contains matches.

bash$ **grep Linux osinfo.txt misc.txt**

osinfo.txt:This is a file containing information about Linux.

osinfo.txt:The GPL governs the distribution of the Linux operating system. misc.txt:The Linux operating system is steadily gaining in popularity.

To force **grep** to show the filename when searching only one target file, simply give /dev/null asthe second file.

$ grep Linux osinfo.txt /dev/null

osinfo.txt:This is a file containing information about Linux.

osinfo.txt:The GPL governs the distribution of the Linux operating system.

If there is a successful match, **grep** returns an exit status of 0, which makes it useful in acondition test in a script, especially in combination with the -q option to suppress output.

```
SUCCESS=0                      # if grep lookup succeedsword=Linux
filename=data.file
grep -q "$word" "$filename" # The "-q" option
                               #+ causes nothing to echo to stdout.
    if [ $? -eq $SUCCESS ]
# if grep -q "$word" "$filename" can replace lines 5 - 7.
then
echo "$word found in $filename"
else
echo "$word not found in $filename"
fi
```

Emulating *grep* in a script

&grp;

How can **grep** search for two (or more) separate patterns? What if you want **grep** to display alllines in a file or files that contain both «pattern1» *and* «pattern2»?

One method is to pipe the result of **grep pattern1** to **grep pattern2**.

For example, given the following file:

```
# Filename: tstfile
This is a sample file.
This is an ordinary text file.
This file does not contain any unusual text.
This file is not unusual.
Here is some text.
```

Now, let's search this file for lines containing *both* «file» and «text» . . .

bash$ **grep file tstfile**

Filename: tstfile This is a sample file.

This is an ordinary text file.

This file does not contain any unusual text.

This file is not unusual.

bash$ grep file tstfile | grep text

This is an ordinary text file.

This file does not contain any unusual text.

Now, for an interesting recreational use of *grep* . . .

egrep -- *extended grep* -- is the same as **grep -E**. This uses a somewhat different, extended set ofRegular Expressions, which can make the search a bit more flexible. It also allows the boolean | (*or*) operator.

bash $ **egrep 'matches | Matches' file.txt**

Line 1 matches.

Line 3 Matches.

Line 4 contains matches, but also Matches

fgrep -- *fast grep* -- is the same as **grep -F**. It does a literal string search (no RegularExpressions), which generally speeds things up a bit.

On some Linux distros, **egrep** and **fgrep** are symbolic links to, or aliases for **grep**, but invoked with the -E and -F options, respectively.

&dictlookup;

agrep (*approximate grep*) extends the capabilities of **grep** to approximate matching. The search string may differ by a specified number of characters from the resulting matches. This utility is not part of the core Linux distribution.

To search compressed files, use **zgrep**, **zegrep**, or **zfgrep**. These also work on non-compressed files, though slower than plain **grep**, **egrep**, **fgrep**. They are handy for searching through a mixed set of files, some compressed, some not.

To search bzipped files, use **bzgrep**.

look

The command **look** works like **grep**, but does a lookup on a «dictionary,» a sorted word list. By default, **look** searches for a match in /usr/dict/words, but a different dictionary file may be specified.

Checking words in a list for validity

&lookup;

sed, awk

Scripting languages especially suited for parsing text files and command output. May be embedded singly or in combination in pipes and shell scripts.

sed

Non-interactive «stream editor», permits using many **ex** commands in batch mode. Itfinds many uses in shell scripts.

awk

Programmable file extractor and formatter, good for manipulating and/or extracting fields(columns) in structured text files. Its syntax is similar to C.

wc

wc gives a «word count» on a file or I/O stream:

bash $ wc /usr/share/doc/sed-4.1.2/README

13 70 447 README

[13 lines 70 words 447 characters]**wc -w** gives only the word count. **wc -l** gives only the line count.

wc -c gives only the byte count.

wc -m gives only the character count.

wc -L gives only the length of the longest line.

Using **wc** to count how many .txt files are in current working directory:

$ ls *.txt | wc -l

Will work as long as none of the "*.txt" files#+ have a linefeed embedded in their name.

Alternative ways of doing this are:

find . -maxdepth 1 -name *.txt -print0 | grep -cz .# (shopt -s nullglob; set -- *.txt; echo $#)

Thanks, S.C.

Using **wc** to total up the size of all the files whose names begin with letters in the range d

- h

bash$ **wc [d-h]* | grep total | awk '{print $3}'**

71832

Using **wc** to count the instances of the word «Linux» in the main source file for this book.bash$ **grep Linux abs-book. sgml | wc -l**

50

Certain commands include some of the functionality of **wc** as options.

... | grep foo | wc -l

This frequently used construct can be more concisely rendered.

... | grep -c foo

Just use the "-c" (or "--count") option of grep.

Thanks, S.C.

Tr command

character translation filter

Must use quoting and/or brackets, as appropriate. Quotes prevent the shell from reinterpretingthe special characters in **tr** command sequences. Brackets should be quoted to prevent expansionby the shell.

Either **tr "A-Z" "*" <filename** or **tr A-Z * <filename** changes all the uppercase letters in filename to asterisks (writes to stdout). On some systems this may not work, but **tr A-Z '[**]'** will.

The -d option deletes a range of characters.

echo "abcdef"   # abcdef

echo "abcdef" | tr -d b-d   # aef

tr -d 0-9 <filename

Deletes all digits from the file "filename".

The --squeeze-repeats (or -s) option deletes all but the first instance of a string of consecutive characters. This option is useful for removing excess whitespace.

bash$ echo "XXXXX" | tr --squeeze-repeats 'X'

X

The -c «complement» option *inverts* the character set to match. With this option, **tr** acts onlyupon those characters *not* matching the specified set.

bash$ echo "acfdeb123" | tr -c b-d +

+c+d+b++++

Note that **tr** recognizes POSIX character classes. bash$ **echo "abcd2ef1" | tr '[:alpha:]' -**

.......2. 1

***toupper*: Transforms a file to all uppercase.**

&ex49;

***lowercase*: Changes all filenames in working directory to lowercase.**

&lowercase;

***du*: DOS to UNIX text file conversion.**

&du;

***rot13*: ultra-weak encryption.**

&rot13

Generating «Crypto-Quote» Puzzles

&cryptoquote;

***tr* variants**

The **tr** utility has two historic variants. The BSD version does not use brackets (**tr a-z A- Z**), but the SysV one does (**tr '[a-z]' '[A-Z]'**). The GNU version of **tr** resembles the BSDone.

fold

A filter that wraps lines of input to a specified width. This is especially useful with the -soption, which breaks lines at word spaces

fmt

Simple-minded file formatter, used as a filter in a pipe to «wrap» long lines of textoutput.

15.26. Formatted file listing.

&ex50;

col

This deceptively named filter removes reverse line feeds from an input stream. It also attempts to replace whitespace with equivalent tabs. The chief use of **col** is in filtering the output from certain text processing utilities, such as **groff** and **tbl**.

column

Column formatter. This filter transforms list-type text output into a «pretty-printed» table by inserting tabs at appropriate places.

Using *column* to format a directory listing

&colm;

colrm

Column removal filter. This removes columns (characters) from a file and writes the file, lacking the range of specified columns, back to stdout. **colrm 2 4 <filename** removes the second through fourth characters from each line of the text file filename.

nl

If the file contains tabs or nonprintable characters, this may cause unpredictable behavior. In such cases, consider using expand and **unexpand** in a pipe preceding **colrm**.

Line numbering filter: **nl filename** lists filename to stdout, but inserts consecutive numbers at the beginning of each non-blank line. If filename omitted, operates on stdin.

The output of **nl** is very similar to **cat -b**, since, by default **nl** does not list blank lines.

15.28. *nl*: A self-numbering script.

&lnum;

pr

Print formatting filter. This will paginate files (or stdout) into sections suitable for hard copy printing or viewing on screen. Various options permit row and columnmanipulation, joining lines, setting margins, numbering lines, adding page headers, and merging files, among other things. The **pr** command combines much of the functionality of **nl**, **paste**, **fold**, **column**, and **expand**.

pr -o 5 --width=65 fileZZZ | more gives a nice paginated listing to screen of fileZZZ with margins set at 5 and 65.

A particularly useful option is -d, forcing double-spacing (same effect as **sed -G**).

SED:

What is sed?

- A non-interactive stream editor
- Interprets sed instructions and performs actions
- Use sed to:

- Automatically perform edits on file(s)
- Simplify doing the same edits on multiple files
- Write conversion programs

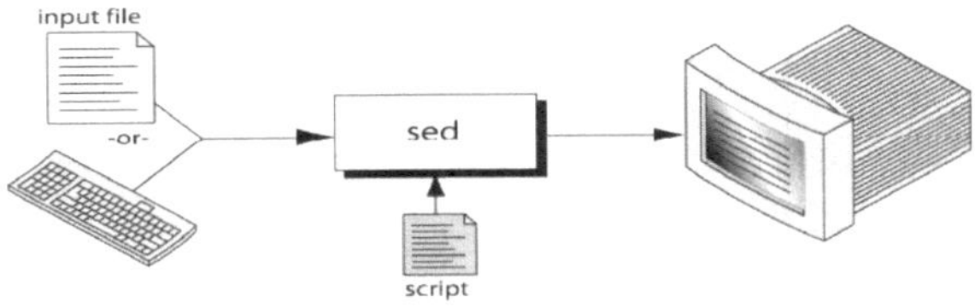

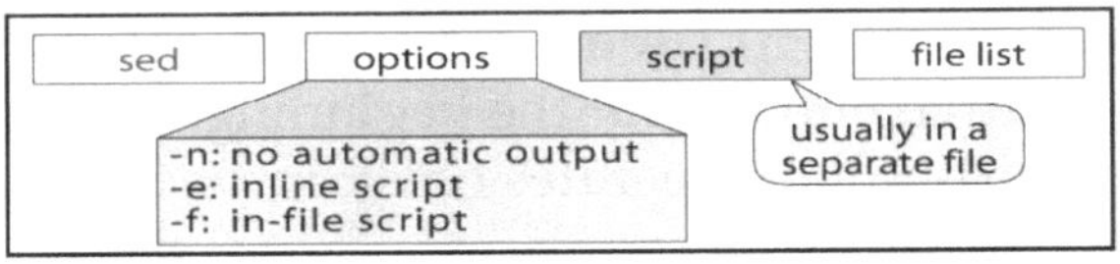

sed command syntax

```
$ sed -e 'address command' input_file
```

(a) Inline Script

```
$ sed -f script.sed input_file
```

(b) Script File

SED OPERATION

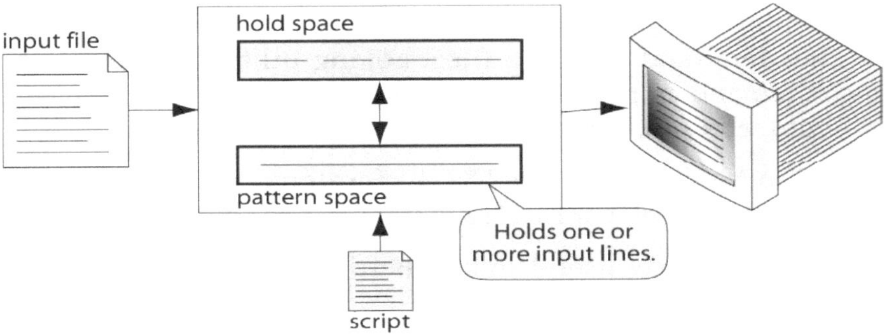

How Does sed Work?

- sed reads line of input
  - line of input is copied into a temporary buffer called pattern space
  - editing commands are applied

- subsequent commands are applied to line in the pattern space, not theoriginal input line
- once finished, line is sent to output(unless –n option was used)

- line is removed from pattern space

- sed reads next line of input, until end of file

Note: input file is unchanged

SED INSTRUCTION FORMAT

- address determines which lines in the input file are to be processed by the command(s)
  - if no address is specified, then the command is applied to each input line
- ADDRESS TYPES:
  - Single-Line address
  - Set-of-Lines address
  - Range address
  - Nested addressSINGLE-LINE ADDRESS
- Specifies only one line in the input file
  - special: dollar sign ($) denotes last line of input file

Examples:

- show only line 3

sed -n -e '3 p' input-file

- show only last line

sed -n -e '$ p' input-file

- substitute "endif" with "fi" on line 10

sed -e '10 s/endif/fi/' input-file

SET-OF-LINES ADDRESS

- use regular expression to match lines
  - written between two slashes
  - process only lines that match

- may match several lines
- lines may or may not be consecutives

Examples:

sed -e '/key/ s/more/other/' input-file
sed -n -e '/r..t/ p' input-file

RANGE ADDRESS

- Defines a set of consecutive lines

Format:

start-addr,end-addr (inclusive)

Examples:

10,50	line-number,line-number
10,/R.E/	line-number,/RegExp/
/R.E./,10	/RegExp/,line-number
/R.E./,/R.E/	/RegExp/,/RegExp/

Example: Range Address

% sed -n -e '/^BEGIN$/,/^END$/p' input-file

- Print lines between BEGIN and END, inclusive

BEGIN

Line 1 of input

Line 2 of input

Line3 of input END

Line 4 of input

Line 5 of input

Nested Address

- Nested address contained within another address

Example:

print blank lines between line 20 and 30

20,30{

/^$/ p

}

Address with !

- address with an exclamation point (!):

  instruction will be applied to all lines that do not match the address

Example:

print lines that do not contain "obsolete"

sed -e '/obsolete/!p' input-file

SED COMMANDS

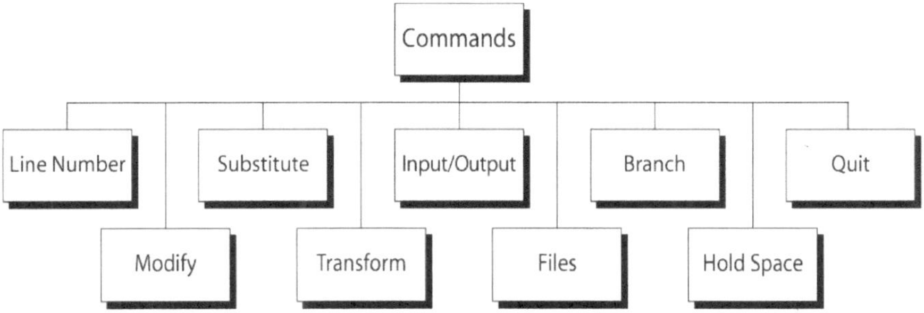

Line Number

- line number command (=) writes the current line number before each matched/output line

Examples:

sed -e '/Two-thirds-time/=' tuition.

datased -e '/^[0-9][0-9]/=' inventory

modify commands

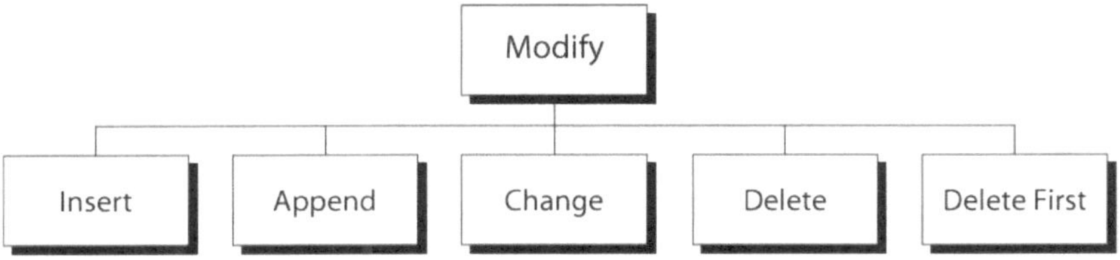

Insert Command: i

- adds one or more lines directly to the output before the address:
  - inserted "text" never appears in sed's pattern space
  - cannot be used with a range address; can only be used with the single-line and set-of-lines address types

Syntax:

[address] i

text

Append Command: a

- adds one or more lines directly to the output after the address:
  - Similar to the insert command (i), append cannot be used with a range address.
  - Appended "text" does not appear in sed's pattern space.

Syntax:

[address] a

text

Change Command: c

- replaces an entire matched line with new text
- accepts four address types:
  - single-line, set-of-line, range, and nested addresses.

Syntax:

[address1[,address2]] c

text

Delete Command: d

- deletes the entire pattern space
  - commands following the delete command are ignored since the deleted text is nolonger in the pattern space

Syntax:

[address1[,address2]] d

Substitute Command (s)

Syntax:

[addr1][,addr2] s/search/replace/[flags]

- replaces text selected by search string with replacement string
- search string can be regular expression
- flags:
  - global (g), i.e. replace all occurrences
  - specific substitution count (integer), default 1

Regular Expressions: use with sed

Metacharacter	Description/Matches...
.	Any one character, except new line
*	Zero or more of preceding character
^	A character at beginning of line
$	A character at end of line
\char	Escape the meaning of *char* following it
[ ]	Any one of the enclosed characters
\(\)	Tags matched characters to be used later
x\{m\}	Repetition of character x, m times
\<	Beginning of word
\>	End of word

Substitution Back References

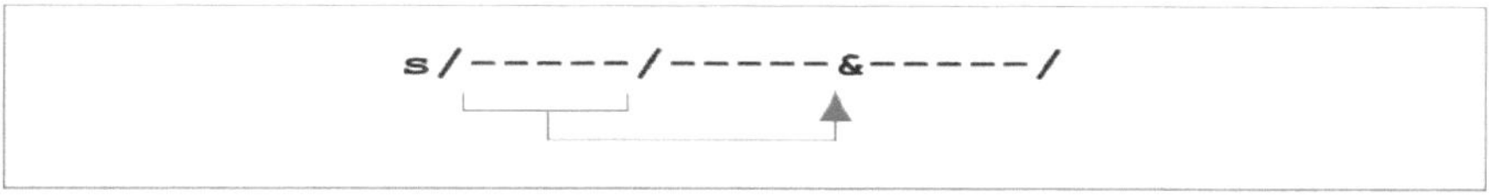

(a) Whole Pattern Substitution

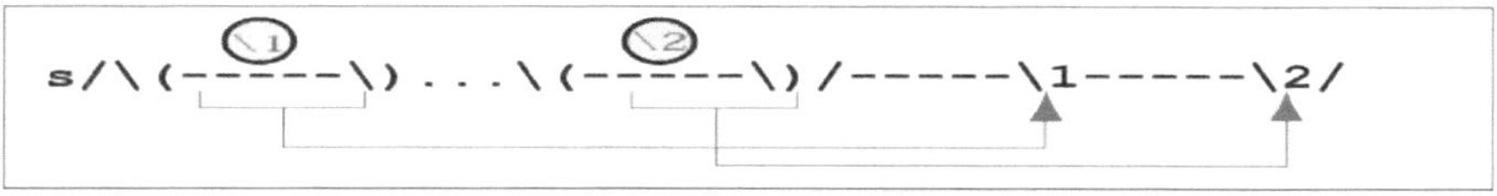

(b) Numbered Buffer Substitution

Example: Replacement String &

$ cat datafile

Charles Main 3.0 .98 3 34

Sharon Gray 5.3 .97 5 23

Patricia Hemenway TB Savage 4.0 .7

4.4 .84 5 4 17

20

AM Main Jr. 5.1 .94 3 13

Margot Weber4.5 .89 5 9

Ann Stephens5.7 .94 5 13

$ sed -e 's/[0-9][0-9]$/&.5/' datafile

Charles Main 3.0 .98 3 34.5

Sharon Gray 5.3 .97 5 23.5

Patricia Hemenway 4.0 .7 4 17.5

TB Savage 4.4 .84 5 20.5

AM Main Jr. 5.1 .94 3 13.5

Margot Weber4.5 .89 5 9

Ann Stephens5.7 .94 5 13.5

Transform Command (y)Syntax:

[addr1][,addr2]y/a/b/

- translates one character 'a' to another 'b'
- cannot use regular expression metacharacters
- cannot indicate a range of characters
- similar to "tr" command

Example:

$ sed -e '1,10y/abcd/wxyz/' datafile

SED I/O COMMANDS

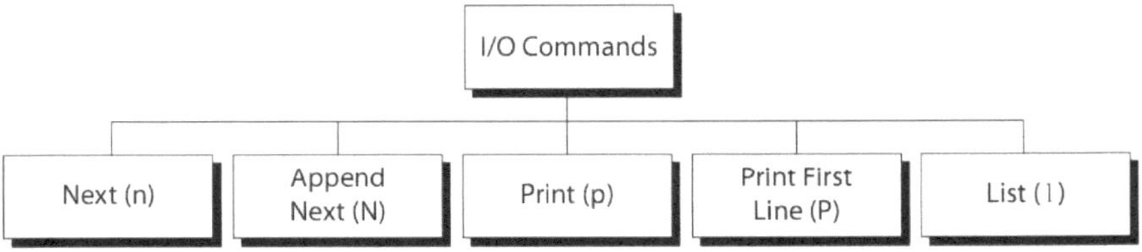

Input (next) Command: n and N

- Forces sed to read the next input line
  - Copies the contents of the pattern space to output
  - Deletes the current line in the pattern space
  - Refills it with the next input line
  - Continue processing
- N (uppercase) Command
  - adds the next input line to the current contents of the pattern space
  - useful when applying patterns to two or more lines at the same timeOutput Command: p and P
- Print Command (p)
  - copies the entire contents of the pattern space to output
  - will print same line twice unless the option "-n" is used
- Print command: P
  - prints only the first line of the pattern space
  - prints the contents of the pattern space up to and including a new line character
  - any text following the first new line is not printed List Command (l)
- The list command: l
  - shows special characters (e.g. tab, etc)
- The octal dump command (od -c) can be used to produce similar resultHold Space
- temporary storage area

  used to save the contents of the pattern space
- 4 commands that can be used to move text back and forth between the pattern space andthe hold space:

File commands

h, H

g, G

- allows to read and write from/ to file while processing standard input
- read: r command
- write: w command

Read File command

Syntax: **r filename**

  - queue the contents of filename to be read and inserted into the output stream at the end of the current cycle, or when the next input line is read

- if filename cannot be read, it is treated as if it were an empty file, withoutany error indication

  - single address only

Write File command

Syntax: **w filename**

  - Write the pattern space to filename
  - The filename will be created (or truncated) before the first input line is read
  - all w commands which refer to the same filename are output through the sameFILE stream

Branch Command (b)

- Change the regular flow of the commands in the script file

Syntax: **[addr1][,addr2]b[label]**

  - Branch (unconditionally) to 'label' or end of script
  - If "label" is supplied, execution resumes at the line following :label; otherwise,control passes to the end of the script

- Branch label

:mylabel

Example: The quit (q) Command

Syntax: **[addr]q**

- Quit (exit sed) when addr is encountered.

Example: Display the first 50 lines and quit

% sed -e '50q' datafile

Same as:

% sed -n -e '1,50p' datafile

% head -50 datafile

AWK

WHAT IS AWK?

- created by: Aho, Weinberger, and Kernighan
- scripting language used for manipulating data and generating reports
- versions of awk
  - awk, nawk, mawk, pgawk, ...
  - GNU awk: gawkWhat can you do with awk?
- awk operation:
  - scans a file line by line
  - splits each input line into fields
  - compares input line/fields to pattern
  - performs action(s) on matched lines
- Useful for:
  - transform data files
  - produce formatted reports
- Programming constructs:

- format output lines
- arithmetic and string operations
- conditionals and loops

The Command:

awk

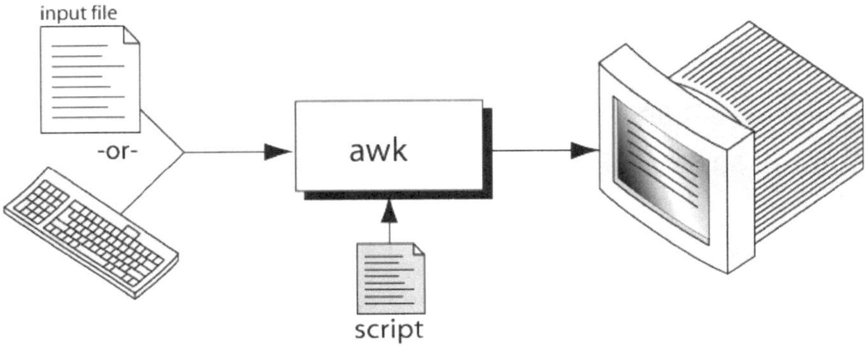

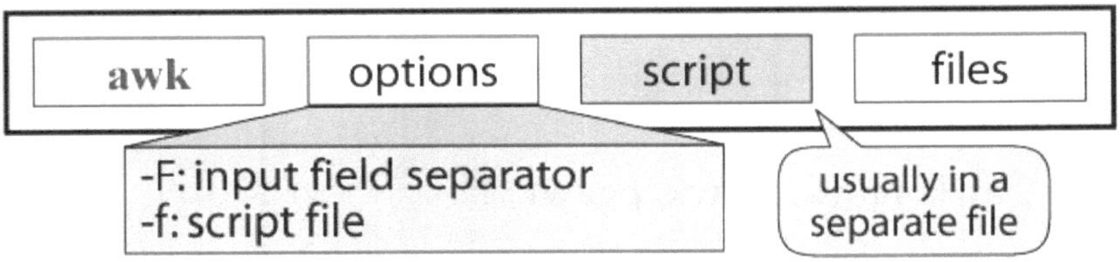

Basic awk Syntax

- **awk [options] 'script' file(s)**
- **awk [options] –f scriptfile file(s)**

Options:

-F to change input field separator

-f to name script fileBasic awk Program

- consists of patterns & actions:

  pattern {action}

  - if pattern is missing, action is applied to all lines
  - if action is missing, the matched line is printed
  - must have either pattern or action

Example:

awk '/for/' testfile

- prints all lines containing string "for" in testfile

BASIC TERMINOLOGY: INPUT FILE

- A field is a unit of data in a line
- Each field is separated from the other fields by the field separator
  - default field separator is whitespace
- A record is the collection of fields in a line
- A data file is made up of records

Example Input File

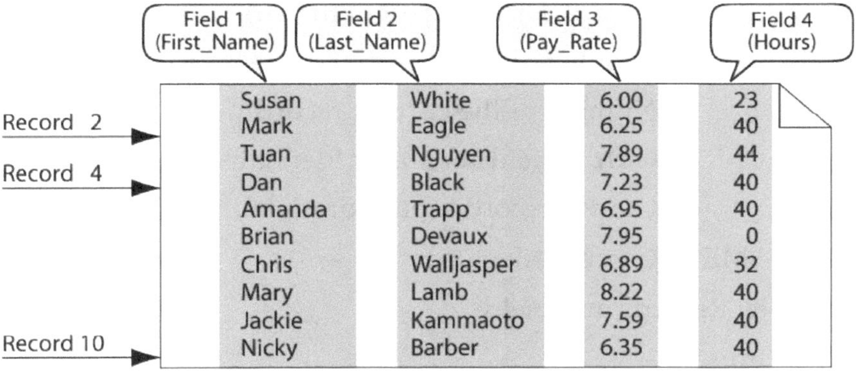

A file with 10 records, each with four fields

Buffers

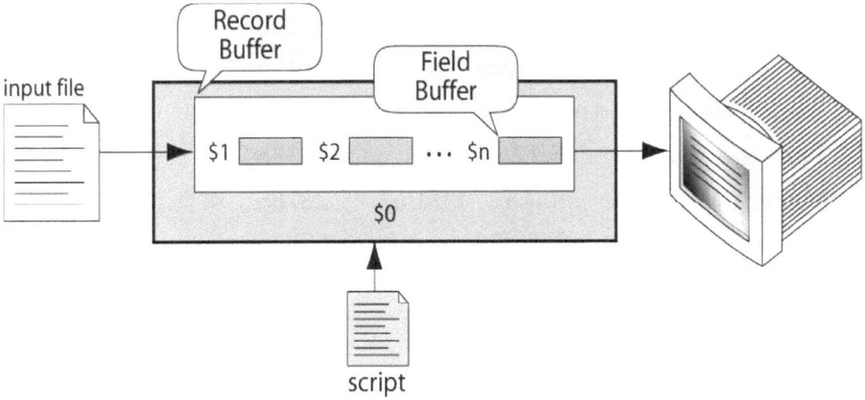

- awk supports two types of buffers:
  record and field
- field buffer:
  - one for each fields in the current record.
  - names: $1, $2, ...
- record buffer :
  - $0 holds the entire recordSome System Variables

FS	Field separator (default=whitespace)
RS	Record separator (default=\n)
NF	Number of fields in current record
NR	Number of the current record
OFS	Output field separator (default=space)
ORS	Output record separator (default=\n)
FILENAME	Current filename

Example: Records and Fields

% cat emps

```
Tom Jones     4424   5/12/66            543354
Mary Adams    5346   11/4/63            28765
Sally Chang   1654   7/22/54            650000
Billy Black   1683   9/23/44            336500
% awk '{print NR, $0}' emps
1 Tom Jones 4424     5/12/66            543354
2 Mary Adams         5346     11/4/63   28765
3 Sally Chang 1654   7/22/54            650000
4 Billy Black 1683   9/23/44            336500
```

Example: Space as Field Separator

% cat emps

Tom Jones	**4424**	**5/12/66 543354**
Mary Adams	**5346**	**11/4/63 28765**
Sally Chang	**1654**	**7/22/54 650000**
Billy Black	**1683**	**9/23/44 336500**

% awk '{print NR, $1, $2, $5}' emps1 Tom Jones 543354

2 Mary Adams 28765

3 Sally Chang 650000

4 Billy Black 336500

Example: Colon as Field Separator

% cat em2

Tom Jones:4424:5/12/66:543354

Mary Adams:5346:11/4/63:28765

Sally Chang:1654:7/22/54:650000

Billy Black:1683:9/23/44:336500

% awk -F: '/Jones/{print $1, $2}' em2

Tom Jones 4424

AWK SCRIPTS

- awk scripts are divided into three major parts:

BEGIN	{Begin's Actions}	Preprocessing
	Pattern {Action} Pattern {Action} Pattern {Action}	Body
END	{End's Actions}	Postprocessing

- comment lines start with #

awk Scripts

- BEGIN: pre-processing
  - performs processing that must be completed before the file processing starts (i.e.,before awk starts reading records from the input file)
  - useful for initialization tasks such as to initialize variables and to create report headings
- BODY: Processing
  - contains main processing logic to be applied to input records
  - like a loop that processes input data one record at a time:
    - if a file contains 100 records, the body will be executed 100 times, one foreach record
- END: post-processing
  - contains logic to be executed after all input data have been processed
  - logic such as printing report grand total should be performed in this part of thescript

Pattern / Action Syntax

```
pattern {statement}
```

(a) One Statement Action

```
pattern {statement1; statement2; statement3}
```

(b) Multiple Statements Separated by Semicolons

```
pattern
{
    statement1
    statement2
    statement3
}
```

(c) Multiple Statements Separated by Newlines

Categories of Patterns

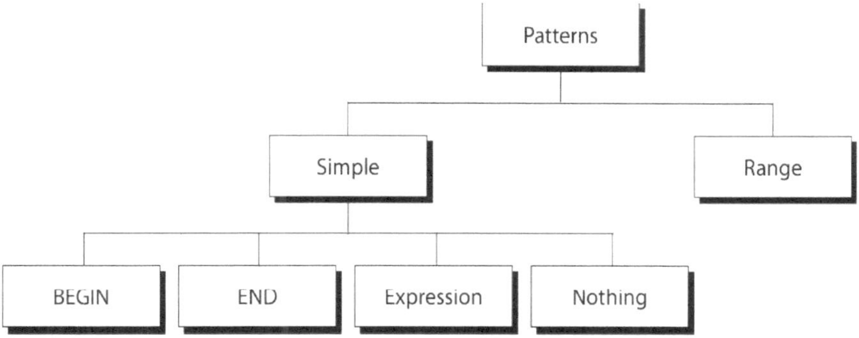

Expression Pattern types

- match
  - entire input record
    regular expression enclosed by '/'s
  - explicit pattern-matching expressions
  ~ (match), !~ (not match)
- expression operators
  - arithmetic
  - relational
  - logical

Example: match input record

% cat employees2

Tom Jones:4424:5/12/66:543354

Mary Adams:5346:11/4/63:28765

Sally Chang:1654:7/22/54:650000

Billy Black:1683:9/23/44:336500

% awk -F: '/00$/' employees2

Sally Chang:1654:7/22/54:650000

Billy Black:1683:9/23/44:336500

Example: explicit match

```
northwest NW    Charles Main      3.0    .983 34
western WE      Sharon Gray 5.3 .97 5 23
southwest SW Lewis Dalsass     2.7 .8      2 18
southern SO     Suan Chin    5.1 .95 4 15
southeast SE    Patricia Hemenway 4.0    .7 4  17
eastern EA      TB Savage    4.4 .84 5    20
northeast NE    AM Main 5.1    .94        3 13
north    NO     Margot Weber   4.5 .89      5    9
central CT      Ann Stephens   5.7 .94      5 13
% awk '$5 ~ /\.[7-9]+/' datafile
southwest SW Lewis Dalsass     2.7 .8      2 18
central CT      Ann Stephens   5.7 .94    5 13
```

Examples: matching with REs

```
% awk '$2 !~ /E/{print $1, $2}' datafile
northwest NW
southwest SW
southern SO
north NO
central CT
% awk '/^[ns]/{print $1}' datafile
northwest
southwest
southern
southeast
northeast
north
```

ARITHMETIC OPERATORS

Operator	Meaning	Example
+	Add	x + y
-	Subtract	x – y
*	Multiply	x * y
/	Divide	x / y
%	Modulus	x % y
^	Exponential	x ^ y

Example:

% awk '$3 * $4 > 500 {print $0}' file

Relational Operators

Operator	Meaning	Example
<	Less than	x < y
< =	Less than or equal	x < = y
==	Equal to	x == y
!=	Not equal to	x != y
>	Greater than	x > y
> =	Greater than or equal to	x > = y
~	Matched by reg exp	x ~ /y/
!~	Not matched by req exp	x !~ /y/

Logical Operators

Operator	**Meaning**	**Example**
&&	Logical AND	a && b
| |	Logical OR	a | | b
!	NOT	! a

Examples:

% awk '($2 > 5) && ($2 <= 15) {print $0}' file

% awk '$3 == 100 | | $4 > 50' file

RANGE PATTERNS

- Matches ranges of consecutive input lines

Syntax:

pattern1 , pattern2 {action}

- pattern can be any simple pattern
- **pattern1** turns action on
- **pattern2** turns action off

Range Pattern Example

AWK ACTIONS

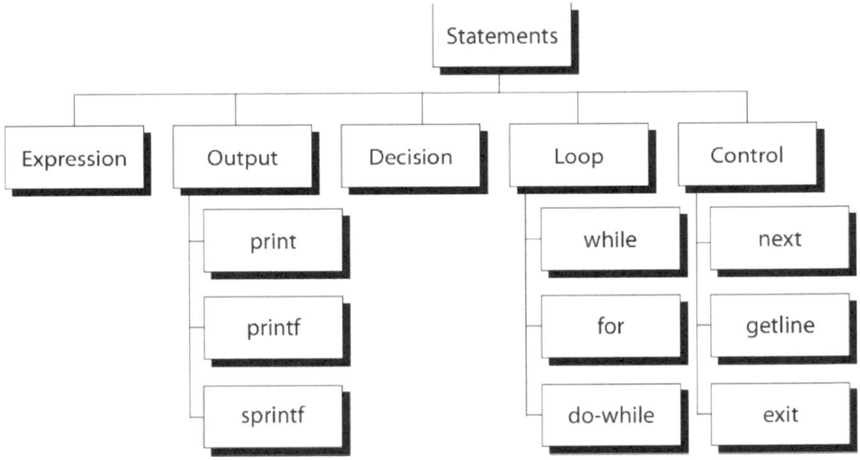

AWK EXPRESSIONS

- Expression is evaluated and returns value
  - consists of any combination of numeric and string constants, variables, operators, functions, and regular expressions
- Can involve variables
  - As part of expression evaluation
  - As target of assignment

awk variables

- A user can define any number of variables within an awk script
- The variables can be numbers, strings, or arrays
- Variable names start with a letter, followed by letters, digits, and underscore
- Variables come into existence the first time they are referenced; therefore, they do notneed to be declared before use
- All variables are initially created as strings and initialized to a null string "" awk Variables

Format:

variable = expression

Examples:

% awk '$1 ~ /Tom/

{wage = $3 * $4; print wage}'
filename

% awk '$4 == "CA" {$4 = "California"; print $0}'
filename

awk assignment operators

= assign result of right-hand-side expression toleft-hand-side variable

++ Add 1 to variable

--	Subtract 1 from variable
+=	Assign result of addition
-=	Assign result of subtraction
*=	Assign result of multiplication
/=	Assign result of division
%=	Assign result of modulo
^=	Assign result of exponentiation

Awk example

- File: grades

  john 85 92 78 94 88

  andrea 89 90 75 90 86

  jasper 84 88 80 92 84

- awk script: average

  # average five grades

  { total = $2 + $3 + $4 + $5 + $6avg = total / 5

  print $1, avg }

- Run as:

  awk –f average grades

Output Statements

print

print easy and simple output

printf

print formatted (similar to C printf)

sprintf

format string (similar to C sprintf)

Function: print

- Writes to standard output

- Output is terminated by ORS
  - default ORS is newline
- If called with no parameter, it will print $0
- Printed parameters are separated by OFS,
  - default OFS is blank
- Print control characters are allowed:
  - \n \f \a \t \\ ...

print example

% awk '{print}' grades

john 85 92 78 94 88

andrea 89 90 75 90 86

% awk '{print $0}' grades

john 85 92 78 94 88

andrea 89 90 75 90 86

% awk '{print($0)}' grades

john 85 92 78 94 88

andrea 89 90 75 90 86

Redirecting print output

- Print output goes to standard outputunless redirected via:

> "file"

>> "file"

| "command"

- will open file or command only once
- subsequent redirections append to already open stream

print Example

% awk '{print $1 , $2 > "file"}' grades

% cat filejohn 85

andrea 89

jasper 84

% awk '{print $1,$2 | "sort"}' grades

andrea 89

jasper 84

john 85

% awk '{print $1,$2 | "sort –k 2"}' grades

jasper 84

john 85

andrea 89

% date

Wed Nov 19 14:40:07 CST 2008

% date |

awk '{print "Month: " $2 "\nYear: ", $6}' Month: Nov

Year: 2008

printf: Formatting output

Syntax:

printf(format-string, var1, var2, ...)

- works like C printf
- each format specifier in "format-string" requires argument of matching type

Format specifiers

%d, %	idecimal integer
%c	single character
%s	string of characters
%f	floating point number
%o	octal number
%x	hexadecimal number
%e	scientific floating point notation
%%	the letter "%"Format specifier examples

Given: x = 'A', y = 15, z = 2.3, and $1 = Bob Smith

Printf Format Specifier	What it Does
%c	*printf("The character is %c \n", x)* output: The character is A
%d	*printf("The boy is %d years old \n", y)* *output: The boy is 15 years old*
%s	*printf("My name is %s \n", $1)* *output: My name is Bob Smith*
%f	*printf("z is %5.3f \n", z)* *output: z is 2.300*

Format specifier modifiers

- between "%" and letter

  %10s

  %7d

  %10.4f

  %-20s

- meaning:
  - width of field, field is printed right justified
  - precision: number of digits after decimal point
  - "-" will left justifysprintf: Formatting text

Syntax:

sprintf(format-string, var1, var2, ...)

- Works like printf, but does not produce output
- Instead it returns formatted string

Example:

```
{
    text = sprintf("1: %d – 2: %d", $1, $2)
    print text
}
```

AWK BUILTIN FUNCTIONS

tolower(string)

- returns a copy of string, with each upper-case character converted to lower-case. Nonalphabetic characters are left unchanged.

  Example: tolower("MiXeD cAsE 123")

  returns "mixed case 123"

toupper(string)

- returns a copy of string, with each lower-case character converted to upper-case.

awk Example: list of products

103:sway bar:49.99

101:propeller:104.99

104:fishing line:0.99

113:premium fish bait:1.00

106:cup holder:2.49

107:cooler:14.89

112:boat cover:120.00

109:transom:199.00

110:pulley:9.88

105:mirror:4.99

108:wheel:49.99

111:lock:31.00

102:trailer hitch:97.95

awk Example: output

Marine Parts R Us

Main catalog

```
Part-id name                    price
=====================================
101     propeller               104.99
102     trailer hitch           97.95
103     sway bar                49.99
104     fishing line            0.99
105     mirror                  4.99
106     cup holder              2.49
107     cooler                  14.89
108     wheel                   49.99
109     transom                 199.00
110     pulley                  9.88
111     lock                    31.00
112     boat cover              120.00
113     premium fish bait       1.00
=====================================

Catalog has 13 parts
```

awk Example: complete

```
BEGIN {
    FS= ":"
    print "Marine Parts R Us"
    print "Main catalog"
    print "Part-id\tname\t\t\t price"
    print "======================================="
}
{
    printf("%3d\t%-20s\t%6.2f\n", $1, $2, $3)count++
}
END {
    print "======================================="
    print "Catalog has " count " parts"
}
```

awk Array

- awk allows one-dimensional array sto store strings or numbers
- index can be number or string
- array need not be declared
    - its size
    - its elements
- array elements are created when first used
    - initialized to 0 or “”

Arrays in awk

Syntax:

arrayName[index] = value

Examples:

list[1] = “one”

list[2] = “three”

list[“other”] = “oh my !”

Illustration: Associative Arrays

- awk arrays can use string as index

Awk builtin split function

split(string, array, fieldsep)

- divides string into pieces separated by fieldsep, and stores the pieces in array
- if the fieldsep is omitted, the value of FS is used.

Example:

split("auto-da-fe", a, "-")

- sets the contents of the array a as follows:

  a[1] = "auto"

  a[2] = "da"

  a[3] = "fe"

Example: process sales data

- input file:

Sales

1	clothing	3141
1	computers	9161
1	textbooks	21312
2	clothing	3252
2	computers	12321
2	supplies	2242
2	textbooks	15462

- output:
  - summary of category sales

Illustration: process each input line

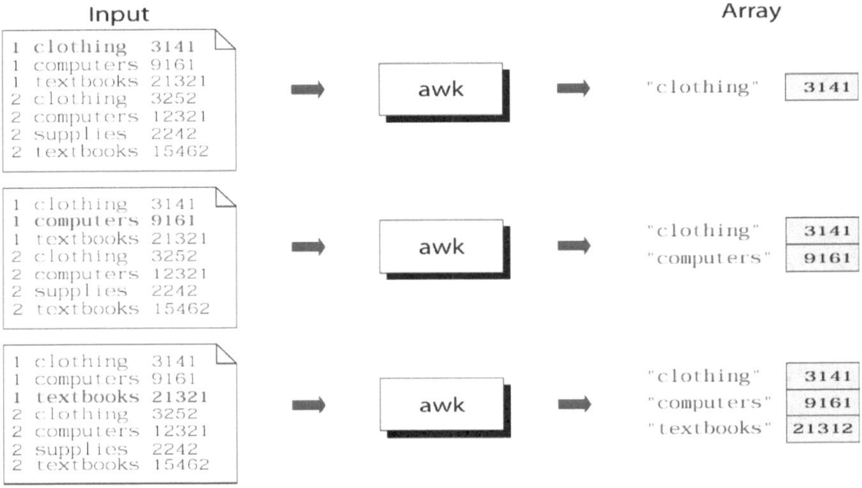

Illustration: process each input line

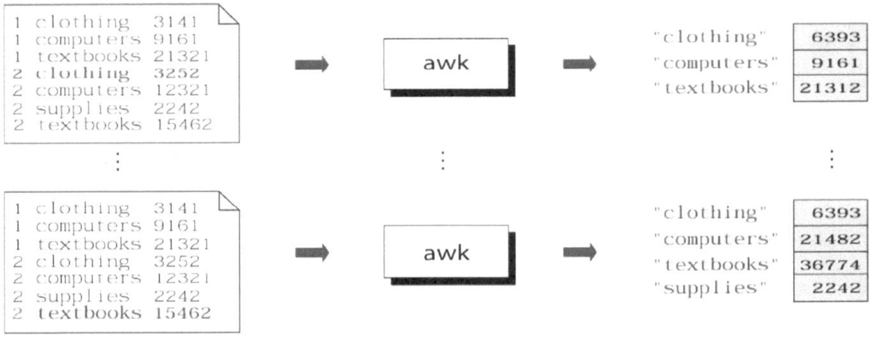

Summary: awk program

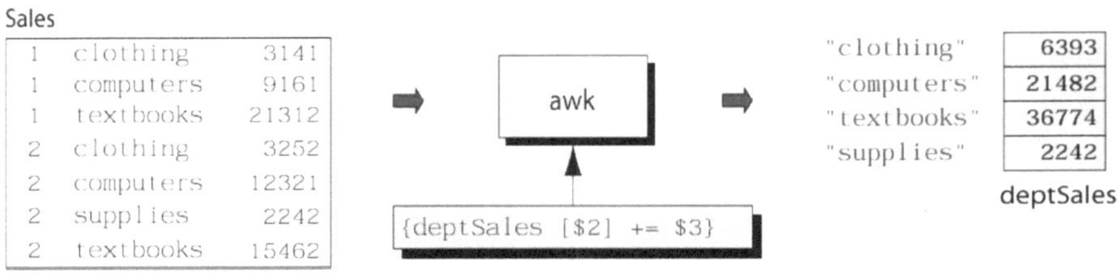

Example: complete program

% **cat sales.awk**

{

```
    deptSales[$2] += $3
}
END {
    for (x in deptSales)
        print x, deptSales[x]
}
% awk –f sales.awk sales
```

Awk control structures

- Conditional
  - if-else
- Repetition
  - for
    - with counter
    - with array index
  - while
  - do-while
  - also: break, continue

if Statement

Syntax:

```
if (conditional expression)
    statement-1
else
    statement-2
```

Example:

```
if ( NR < 3 )
    print $2
else
    print $3
```

for Loop

Syntax:

```
for (initialization; limit-test; update)
    statement
```

Example:

```
for (i = 1; i <= NR; i++)
{
    total += $i
    count++
}
```

for Loop for arrays

Syntax:

```
for (var in array)
    statement
```

Example:

```
for (x in deptSales)
{
    print x, deptSales[x]
}
```

while Loop

Syntax:

```
while (logical expression)
    statement
```

Example:

```
i = 1
while (i <= NF)
{
```

```
    print i, $
    ii++
}
```

do-while Loop

Syntax:

```
    do
        statement
    while (condition)
```

- statement is executed at least once, even if condition is false at the beginning

Example:

```
    i = 1
    do {
    print $0i++
    } while (i <= 10)
```

loop control statements

- break

UNIT : II

Working with the Bourne again Shell (Bash)

WORKING WITH BOURNE SHELL

- The **Bourne shell**, or **sh**, was the default Unix shell of Unix Version 7. It was developedby Stephen Bourne, of AT&T Bell Laboratories.
- A **Unix shell**, also called "the command line", provides the traditional user interface forthe Unix operating system and for Unix-like systems. Users direct the operation of the computer by entering command input as text for a shell to execute.

There are many different shells in use. They are

- Bourne shell (**sh**)
- C shell (**csh**)
- Korn shell (ksh)

Bourne Again shell (bash)

- When we issue a command the shell is the first agency to acquire the information. Itaccepts and interprets user requests. The shell examines &rebuilds the commands &leaves the execution work to kernel. The kernel handles the h/w on behalf of thesecommands &all processes in the system.

- The shell is generally sleeping. It wakes up when an input is keyed in at the prompt. ThisINPUT IS ACTUALLY INPUT TO THE PROGRAM THAT REPRESENTS THE SHELL.

SHELL RESPONSIBILITIES

- Program Execution
- Variable and Filename Substitution
- I/O Redirection
- Pipeline Hookup
- Environment Control
- Interpreted Programming Language

Program Execution:

- The shell is responsible for the execution of all programs that you request from yourterminal.
- Each time you type in a line to the shell, the shell analyzes the line and then determineswhat to do.
- The line that is typed to the shell is known more formally as the command line. The shellscans this command line and determines the name of the program to be executed and what arguments to pass to the program.

Variable and Filename Substitution:

- Like any other programming language, the shell lets you assign values to variables. Whenever you specify one of these variables on the command line, preceded by a dollar sign, the shell substitutes the value assigned to the variable at that point.

I/O Redirection:

- It is the shell's responsibility to take care of input and output redirection on the commandline. It scans the command line for the occurrence of the special redirection characters <, >, or >>

Pipeline Hookup:

- Just as the shell scans the command line looking for redirection characters, it also looks for the pipe character |. For each such character that it finds, it connects the standard output from the command preceding the | to the standard input of the one following the |. It then initiates execution of both programs.

Environment Control:

- The shell provides certain commands that let you customize your environment. Your environment includes home directory, the characters that the shell displays to prompt you to type in a command, and a list of the directories to be searched whenever you request that a program be executed.

Interpreted Programming Language:

- The shell has its own built-in programming language. This language is interpreted, meaning that the shell analyzes each statement in the language one line at a time and then executes it. This differs from programming languages such as C and FORTRAN, in which the programming statements are typically compiled into a machine-executable form before they are executed.
- Programs developed in interpreted programming languages are typically easier to debug and modify than compiled ones. However, they usually take much longer to execute thantheir compiled equivalents.

PIPES

- Standard I/p & standard o/p constitute two separate streams that can be individually manipulated by the shell. The shell connects these streams so that one command takes I /p from other using **pipes**.

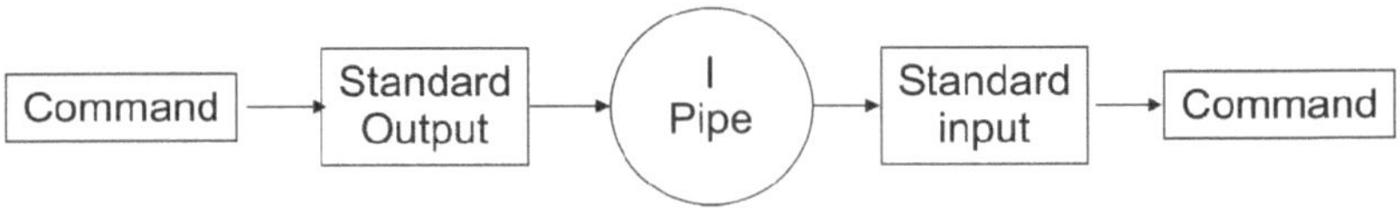

Who produces the list of users , to save this o/p in a file use

$who > user.lst

To count the no. of lines in this user.lst use

$wc –l <user.lst

- This method of using two commands in sequence has certain disadvantages.

1. The process is slow.
2. An intermediate file is required that has to be removed after the command has completed itsrun.
3. When handling large files, temporary files can build up easily &eat up disk space in no time.

- Instead of using two separate commands, the shell can use a special operator as the connector of two commands-the pipe(|).

$who | wc -l

- Here who is said to be piped to wc.
- When a sequence of commands is combined together in this way a pipeline is said to beformed.
- To count no. of files in current directory

$ls | wc –l

- There's no restriction on the no. of commands u can use a pipeline.

REDIRECTION

- Many of the commands that we used sent their output to the terminal and also taking the input from the keyboard. These commands are designed that way to accept not only fixedsources and destinations. They are actually designed to use a character stream without knowing its source and destination.

- A stream is a sequence of bytes that many commands see as input and output. Unix treatsthese streams as files and a group of unix commands reads from and writes to these files.
- There are 3 streams or standard files. The shell sets up these 3 standard files and attachesthem to user terminal at the time of logging in.
- Standard i/p----------default source is the keyboard.
- Standard o/p----------default source is the terminal.
- Standard error----------default source is the terminal.
- Instead of input coming from the keyboard and output and error going to the terminal,they can be **redirected** to come from or go to any file or some other device.

Standard o/p: It has 3 sources.

- The terminal, default source
- A file using redirection with >, >>
- Another program using a pipeline.
- Using the symbols >,>> u can redirect the o/p of a command to a file.

$who> newfile

- If the output file does not exist the shell creates it before executing the command. If itexists the shell overwrites it.
- $who>> newfile

STANDARD I/P:

- The keyboard, default source
- A file using redirection with <
- Another program using a pipeline.
- $wc < calc.lst or
- $wc calc.lst or $wc

STANDARD ERROR:

- When u enter an incorrect command or trying to open a non existing file, certain diagnostic messages show up on the screen. This is the standard error stream.
- Trying to cat a nonexistent file produces the error stream.

$cat bar

Cat: cannot open bar :no such file or directory

- The standard error stream can also be redirected to a file.

$cat bar 2> errorfile

Here 2 is the file descriptor for standard error file.

Each of the standard files has a number called a file descriptor, which is used for identification.

0—standard i/p

1---standard o/p

2---standard error

HERE DOCUMENTS

There are occasions when the data of ur program reads is fixed & fairly limited.

- The shell uses << symbols to read data from the same file containing the script. This referred to as a here document, signifying that the data is here rather than in a separatefile.
- Any command using standard i/p can also take i/p from a here document.
- This feature is useful when used with commands that don't accept a file name asargument.
- Example:

mail Juliet << MARK

Ur pgm for printing the invoices has been executed on `date`. Check the print queue

The updated file is known as $flname

MARK

- The shell treats every line followed by three lines of data and a delimited by MARK as input to the command. Juliet at other end will only see the three lines of message text, the word MARK itself doesn't show up.
- Shell metacharacters
- The shell consists of large no. of metacharacters. These characters plays vital role in Unixprogramming.

Executing a shell script

- you can execute a shell script by invoking its filename.
- $filename
- U can also use *sh* command which takes script name as argument.
- $sh filename

TYPES OF METACHARACTERS:

- 1.File substitution
- 2.I/O redirection
- 3.Process execution
- 4.Quoting metacharacters
- 5.Positional parameters
- 6.Special characters

Filename substitution:

- These metacharacters are used to match the filenames in a directory.
- Metacharacter significance
- * matches any no. of characters
- ? matches a single character
- [ijk] matches a single character either i,j,k

- [!ijk]   matches a single character that is notan I,j,k

I/O redirection:

- These special characters specify from where to take i/p & where to send o/p.
- >- to send the o/p to a specific file
- < - to take i/p from specific location but not from keyboard.
- >>- to save the o/p in a particular file at the end of that file without overwriting it.
- <<- to take i/p from standard i/p file.

Process execution:

- ; -is used when u want to execute more then one command at $ prompt.
- Eg:$date; cat f1>f2
- () –used to group the commands.
- Eg:(date; cat f1) >f2
- & -used to execute the commands in background mode.
- Eg: $ls &
- && -this is used when u want to execute the second command only if the first commandexecuted successfully.
- Eg:$grep Unix f1 && echo Unix found
- $cc f1 && a.out
- | | - used to execute the second command if first command fails.
- Eg:$grep unix f1 | | echo no unix

Quoting:

- \ (backslash)- negates the special property of the single character following it.
- Eg:$echo \? * \?

- ?*?
- ' '(pair of single quotes)-negates the special properties of all enclosed characters.
- Eg:$echo 'send $100 to whom?'
- " "(pair of double quotes)-negates the special properties of all enclosed characters except $,`,\ .
- Eg:$echo "today date is $date" or
- $echo "today date is `date` "

Positional parameters:

- $0- gives the name of the command which is being executed.
- $*-gives the list of arguments.
- $#-gives no. of arguments.

Special parameters:

- $$- gives PID of the current shell.
- $?-gives the exit status of the last executed command.
- $!-gives the PID of last background process.
- $- -gives the current setting of shell.

SHELL VARIABLES

- U can define & use variables both in the command line and shell scripts. These variablesare called shell variables.
- No type declaration is necessary before u can use a shell variable.
- Variables provide the ability to store and manipulate the information with in the shell program. The variables are completely under the control of user.
- Variables in Unix are of two types.

  1.Environmental variables or system variables

  2.User defined variables

User-defined variables:

Generalized form:

variable=value.

Eg: $x=10

$echo $x

10

To remove a variable use unset.

$unset x

- All shell variables are initialized to null strings by default. To explicitly set null valuesuse

x= or x='' or x=""

- To assign multiword strings to a variable use

$msg='u have a mail'

Environment Variables

- They are initialized when the shell script starts and normally capitalized to distinguishthem from user-defined variables in scripts
- To display all variables in the local shell and their values, type the **set** command
- The **unset** command removes the variable from the current shell and sub shell

Environment Variables	Description
$HOME	Home directory
$PATH	List of directories to search for commands
$PS1	Command prompt
$PS2	Secondary prompt
$SHELL	Current login shell
$0	Name of the shell script
$#	No . of parameters passed
$$	Process ID of the shell script

Parameter Variables Parameter Variable

$1, $2, ... The parameters given to the script

$*--→ A list of all the parameters separated by the first character of IFS

$@ --- >A list of all the parameters that doesn't use the IFS environment variable

Shell commands

read:

- The read statement is a tool for taking input from the user i.e. making scripts interactive. It is used with o Input supplied through the standard input is read into these variables.

$read name

What ever you u entered is stored in the variable name.

printf:

Printf is used to print formatted o/p.

printf "format" arg1 arg2 ...

Eg:

$ printf "This is a number: %d\n" 10

This is a number: 10

$

Printf supports conversion specification characters like %d, %s ,%x ,%o....

Exit status of a command:

- Every command returns a value after execution .
- This value is called the exit status or return value of a command.
- This value is said to be true if the command executes successfully and false if it fails. There is special parameter used by the shell it is the $?. It stores the exit status of a command

exit:

- The exit statement is used to prematurely terminate a program.
- When this statement is encountered in a script, execution is halted and control is ----- returned to the calling program-in most cases the shell.
- U don't need to place exit at the end of every shell script because the shell knows -----when script execution is complete.

set:

- Set is used to produce the list of currently defined variables.

$set

- Set is used to assign values to the positional parameters.

$set welcome to Unix

The do-nothing(:)Command

- It is a null command.
- In some older shell scripts, colon was used at the start of a line to introduce a comment,but modern scripts uses # now.

expr:

- The **expr** command evaluates its arguments as an expression:
- $ expr 8 + 6
- **14**
- **$ x=`expr 12 / 4 `**
- **$ echo $x**
- 3

export:

- There is a way to make the value of a variable known to a sub shell, and that's byexporting it with the export command. The format of this command is

export variables

where variables is the list of variable names that you want exported. For any sub shells that getexecuted from that point on, the value of the exported variables will be passed down to the subshell.

eval:

- eval scans the command line twice before executing it. General form for eval iseval command-line

Eg:

$ cat last

eval echo \$$#

$ last one two three fourfour

${n}

If u supply more than nine arguments to a program, u cannot access the tenth and greaterarguments with $10, $11, and so on.

${n} must be used. So to directly access argument 10, you must write

${10}

Shift command:

The shift command allows u to effectively left shift your positional parameters. If u execute thecommand

Shift

What ever was previously stored inside $2 will be assigned to $1, whatever was previouslystored in $3 will be assigned to $2, and so on. The old value of $1 will be irretrievably lost.

File Conditions

-d file True if the file is a directory

-e file True if the file exists

-f file True if the file is a regular file

-g file True if set-group-id is set on file

-r file True if the file is readable

-s file True if the file has non-zero size

-u file True if set-user-id is set on file

-w file True if the file is writeable

-x file True if the file is executable

Example

```
$ mkdir temp
$ if [ -f temp ]; then
>   echo "temp is a directory"
>   fi
```

Arithmetic Comparison

ep1* -eq *ep2 True if ***ep1* = *ep2***

ep1* -ne *ep2 True if ***ep1* ^{1} *ep2***

ep1* -gt *ep2 True if ***ep1* > *ep2***

ep1* -ge *ep2 True if ***ep1* >= *ep2***

ep1* -lt *ep2 True if ***ep1* < *ep2***

ep1* -le *ep2 True if ***ep1* <= *ep2***

! ***ep*** True if ***ep*** is false

Example

```
$ x=5; y=7
$ if [ $x -lt $y ]; then
>   echo "x is less than y"
>   fi
```

BASIC SHELL PROGRAMMING

- A script is a file that contains shell commandsdata structure: variables
- control structure: sequence, decision, loopShebang line for bashshell script:
- #! /bin/bash #!

/bin/sh

- to run:
- make executable: **% chmod +x script** invoke via:% **./script**
BASH SHELL PROGRAMMING

Input prompting user

command line arguments Decision:

if-then-else

Repetition

- Repetition
- do-while, repeat-until
- for
- select
- Functions
- TrapsUser input
- shell allows to prompt for user input

Syntax:

read varname [more vars]

- or

read –p "prompt" varname [more vars]

- words entered by user are assigned to

varname and "**more vars**"

- last variable gets rest of input line

User input example

#! /bin/sh

read -p "enter your name: " first lastecho "First name: $first"

echo "Last name: $last"Special shell variables

Parameter	Meaning
$0	Name of the current shell script
$1-$9	Positional parameters 1 through 9
$#	The number of positional parameters
$*	All positional parameters, "$*" is one string
$@	All positional parameters, "$@" is a set of strings
$?	Return status of most recently executed command
$$	Process id of current process

Examples: Command Line Arguments

% set tim bill ann fred

$1 $2 $3 $4

% echo $*

tim bill ann fred

% echo $#4

% echo $1tim

% echo $3 $4ann fred

The 'set' command can be used to assign values to positional parameters

bash control structures

- if-then-else
- case
- loops
  - for
  - while
  - until

select

if statement

if commandthen

statements

fi

- statements are executed only if **command** succeeds, i.e. has return status "0"

test commandSyntax:

test expression[expression]

- evaluates 'expression' and returns true or falseExample:

if test –w "$1" then

echo "file $1 is write-able" fi

The simple if statement

if [condition]; then

statements

fi

- executes the statements only if **condition** is true

The if-then-else statement

if [condition]; then

statements-1

else

statements-2fi

- executes statements-1 if condition is true
- executes statements-2 if condition is falseThe if...statement

if [condition]; then

statements elif [condition]; then

statementelse

statements

fi

- The word **elif** stands for "else if"
- It is part of the if statement and cannot be used by itself

Relational Operators

Meaning	Numeric	String
Greater than	-gt	
Greater than or equal	-ge	
Less than	-lt	
Less than or equal	-le	
Equal	-eg	= or ==
Not equal	-ne	!=
str1 is less than str2		str1 < str2
str1 is greater str2		str1 > str2
String length is greater thanzero		-n str
String length is zero		-z str

Logical operators

! not

&& and

| | or

Example: Using the ! Operator

```
#!/bin/bash
read -p "Enter years of work: " Yearsif [ ! "$Years" -lt 20 ]; then
echo "You can retire now."else
echo "You need 20+ years to retire"
fi
```

Example: Using the && Operator

#!/bin/bashBonus=500

read -p "Enter Status: " Statusread -p "Enter Shift: " Shift

if [["$Status" = "H" && "$Shift" = 3]]then

echo "shift $Shift gets \$$Bonus bonus"else

echo "only hourly workers in"echo "shift 3 get a bonus"

fi

Example: Using the | | Operator

#!/bin/bash

read -p "Enter calls handled:" CHandleread -p "Enter calls closed: " CClose

if [["$CHandle" -gt 150 | | "$CClose" -gt 50]]then

echo "You are entitled to a bonus"else

echo "You get a bonus if the calls"echo "handled exceeds 150 or" echo "calls closed exceeds 50"

fi

File Testing

Meaning

-d file True if 'file' is a directory

-f file True if 'file' is an ord. file

-r file True if 'file' is readable

-w file True if 'file' is writable

-x file True if 'file' is executable

-s file True if length of 'file' is nonzero

Example: File Testing

#!/bin/bash

echo "Enter a filename: "read filename

if [! –r "$filename"]then

echo "File is not read-able"exit 1

fi

Example: File Testing

#! /bin/bash

if [$# -lt 1]; then

echo "Usage: filetest filename"exit 1

fi

if [[! -f "$1" | | ! -r "$1" | | ! -w "$1"]]then

echo "File $1 is not accessible"

exit 1fi

Example: if... Statement

The following THREE *if*-conditions produce the same result

* DOUBLE SQUARE BRACKETS

read -p "Do you want to continue?" replyif [[$reply = "y"]]; then

echo "You entered " $reply

fi

* SINGLE SQUARE BRACKETS

read -p "Do you want to continue?" replyif [$reply = "y"]; then

echo "You entered " $reply

fi

* "TEST" COMMAND

read -p "Do you want to continue?" replyif test $reply = "y"; then

echo "You entered " $reply

fi

Example: if..elif... Statement

#!/bin/bash

read -p "Enter Income Amount: " Income read -p "Enter Expenses Amount: " Expenselet Net=$Income-$Expense

if ["$Net" -eq "0"]; then

echo "Income and Expenses are equal - breakeven."elif ["$Net" -gt "0"]; then

echo "Profit of: " $Netelse

echo "Loss of: " $Net

fi

The case Statement

- use the case statement for a decision that is based on multiple choicesSyntax:

case word in

pattern1) command-list1

;;

pattern2) command-list2

;;

patternN) command-listN

;;

esac

case pattern

- checked against word for match
- may also contain:

*

?

[…] [:class:]

- multiple patterns can be listed via:

|

Example 1: The case Statement

#!/bin/bash

echo "Enter Y to see all files including hidden files"echo "Enter N to see all non-hidden files"

echo "Enter q to quit"

read -p "Enter your choice: " replycase $reply in

Y | YES) echo "Displaying all (really…) files"ls -a ;;

N | NO) echo "Display all non-hidden files..."ls ;;

Q) exit 0 ;;

*) echo "Invalid choice!"; exit 1 ;;esac

Example 2: The case Statement#!/bin/bash

ChildRate=3 AdultRate=10SeniorRate=7

read -p "Enter your age: " agecase $age in

[1-9] | [1][0-2]) # child, if age 12 and youngerecho "your rate is" '$'"$ChildRate.00" ;;

adult, if age is between 13 and 59 inclusive

[1][3-9] | [2-5][0-9])

echo "your rate is" '$'"$AdultRate.00" ;;[6-9][0-9])
senior, if age is 60+

echo "your rate is" '$'"$SeniorRate.00" ;;esac

Bash programming: so far

- Data structure
  - Variables
  - Numeric variables
  - Arrays
- User input
- Control structures
  - if-then-else
  - case

Bash programming: still to come

- Control structures
- Repetition
- do-while, repeat-until
- for
- select
- Functions
- Trapping signals

Repetition Constructs

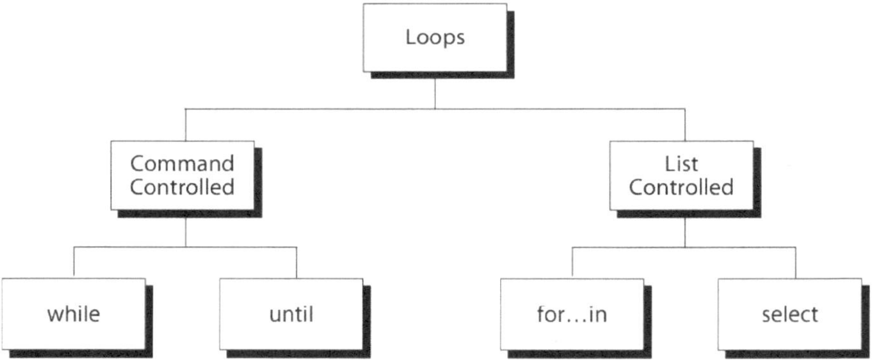

The while Loop

- Purpose:

To execute commands in "command-list" as long as "expression" evaluates to trueSyntax:

while [expression]do

command-list

done

Example: Using the while Loop

#!/bin/bash COUNTER=0

while [$COUNTER -lt 10]do

echo The counter is $COUNTERlet COUNTER=$COUNTER+1

done

Example: Using the while Loop

#!/bin/bashCont="Y"

while [$Cont = "Y"]; dops -A

read -p "want to continue? (Y/N)" reply Cont=`echo $reply | tr [:lower:] [:upper:]` done

echo "done"

Example: Using the while Loop

#!/bin/bash

copies files from home- into the webserver- directory# A new directory is created every hour PICSDIR=/home/carol/pics WEBDIR=/var/www/carol/webcam

while true; do

DATE=`date +%Y%m%d`HOUR=`date +%H`

mkdir $WEBDIR/"$DATE" while [$HOUR -ne "00"]; do

DESTDIR=$WEBDIR/"$DATE"/"$HOUR"

mkdir "$DESTDIR"

mv $PICSDIR/*.jpg "$DESTDIR"/sleep 3600

HOUR=`date +%H`done

done

The until Loop

- Purpose:

To execute commands in "command-list" as long as "expression" evaluates to false

Syntax:

until [expression]do

command-list

done

Example: Using the until Loop

#!/bin/bash COUNTER=20

until [$COUNTER -lt 10]do

echo $COUNTERlet COUNTER-=1

done

Example: Using the until Loop

#!/bin/bashStop="N"

until [$Stop = "Y"]; dops -A

read -p "want to stop? (Y/N)" reply

Stop=`echo $reply | tr [:lower:] [:upper:]`done

echo "done"

The for Loop

- Purpose:

To execute commands as many times as the number of words in the "argument-list"Syntax:

for variable in argument-listdo

commands

done

Example 1: The for Loop

#!/bin/bash

for i in 7 9 2 3 4 5do

echo $idone

Example 2: Using the for Loop

#!/bin/bash

compute the average weekly temperaturefor num in 1 2 3 4 5 6 7

do

read -p "Enter temp for day $num: " Temp

let TempTotal=$TempTotal+$Tempdone

let AvgTemp=$TempTotal/7

echo "Average temperature: " $AvgTemplooping over arguments

- simplest form will iterate over all command line arguments:

#! /bin/bashfor parm do

echo $parmdone

Select command

- Constructs simple menu from word list
- Allows user to enter a number instead of a word
- User enters sequence number corresponding to the word

Syntax:

select WORD in LISTdo

RESPECTIVE-COMMANDS

done

- Loops until end of input, i.e. ^d (or ^c)

Select example#! /bin/bash

select var in alpha beta gammado

echo $vardone

- Prints:
-) alpha
- **2) beta**
- 3) gamma
- **#? 2**
- beta

- **#? 4**
- #? 1
- **alpha**

Select detail

- PS3 is select sub-prompt
- $REPLY is user input (the number)

#! /bin/bash

PS3="select entry or ^D: "select var in alpha beta

do

echo "$REPLY = $var" done

Output:select ...

1) **alpha**
2) beta

? 2

2 = beta

? 1

1 = alpha

#!/bin/bash

echo "script to make files private" echo "Select file to protect:" select FILENAME in *

do

echo "You picked $FILENAME ($REPLY)" chmod go-rwx "$FILENAME"

echo "it is now private" done

Output:

select ...

1) alpha

2) beta

? 2

2 = beta

? 1

1 = alpha

break and continue

- Interrupt for, while or until loop
- The break statement
  - transfer control to the statement AFTER the done statement
  - terminate execution of the loop
- The continue statement
  - transfer control to the statement TO the done statement
  - skip the test statements for the current iteration
  - continues execution of the loopThe break command

while [condition]do

cmd-1breakcmd-n

done

echo "done"

The continue command

while [condition]do

cmd-1 continuecmd-n

done

echo "done"

Example:

for index in 1 2 3 4 5 6 7 8 9 10do

if [$index –le 3]; thenecho "continue" continue

fi

echo $index

if [$index –ge 8]; thenecho "break"

break

fi

done

Bash shell programming

- Sequence
- Decision:
  - if-then-else
  - case
- Repetition
  - do-while, repeat-until
  - for
  - select
- Functions
- Traps

Shell Functions

- A shell function is similar to a shell script
  - stores a series of commands for execution later
  - shell stores functions in memory
  - shell executes a shell function in the same shell that called it
- Where to define
  - In .profile
  - In your script
  - Or on the command line
- Remove a function

- Use unset built-in
  - must be defined before they can be referenced
  - usually placed at the beginning of the scriptSyntax:

function-name () {statements

}

Example: function

#!/bin/bash

funky () {

This is a simple function echo "This is a funky function."

echo "Now exiting funky function."

}

declaration must precede call:funky

Example: function

#!/bin/bash

fun () { # A somewhat more complex function.

JUST_A_SECOND=1

let i=0 REPEATS=30

echo "And now the fun really begins." while [$i -lt $REPEATS]

do

}

fun

done

echo "-------FUNCTIONS are fun
>"

sleep $JUST_A_SECONDlet i+=1

Function parameters

- Need not be declared

- Arguments provided via function call are accessible inside function as $1, $2, $3, …

$# reflects number of parameters

$0 still contains name of script

(not name of function) Example: function with parameter #! /bin/sh

```
testfile() {
if [ $# -gt 0 ]; then
if [[ -f $1 && -r $1 ]]; thenecho $1 is a readable file
else
echo $1 is not a readable file
fi
fi
}
testfile . testfile funtest
```

Example: function with parameters

```
#! /bin/bash checkfile() {
for filedo
if [ -f "$file" ]; then echo "$file is a file"
else
if [ -d "$file" ]; then
echo "$file is a directory"
fi
fi
done
}
checkfile . funtest
```

Local Variables in Functions

- Variables defined within functions are global,

o i.e. their values are known throughout the entire shell program

- keyword "local" inside a function definition makes referenced variables "local" to thatfunction

Example: function

#! /bin/bash

global="pretty good variable"foo () {

local inside="not so good variable"echo $global

echo $inside global="better variable"

}

echo $globalfoo

echo $globalecho $inside

Handling signals

- Unix allows you to send a signal to any process
- -1 = hangup **kill -HUP 1234**
- -2 = interrupt with ^C **kill -2 1235**
- no argument = terminate
kill 1235
- -9 = kill **kill -9 1236**
- -9 cannot be blocked
- list your processes with

ps -u userid

Signals on Linux

% **kill -l**

Shell script examples

Example#!/bin/sh

```
echo "Is it morning? (Answer yes or no)"read timeofday
if [ $timeofday = "yes" ]; thenecho "Good Morning"
else
fi exit 0
echo "Good afternoon"
elif - Doing further Checks#!/bin/sh
echo "Is it morning? Please answer yes or no"read timeofday
if [ $timeofday = "yes" ]; thenecho "Good Morning"
elif [ $timeofday = "no" ]; thenecho "Good afternoon"
else
echo "Wrong answer! Enter yes or no"exit 1
fi exit 0
elif - Doing further Checks#!/bin/sh
echo "Is it morning? Please answer yes or no"read timeofday
if [ $timeofday = "yes" ]; thenecho "Good Morning"
elif [ $timeofday = "no" ]; thenecho "Good afternoon"
else
fi exit 0
echo "Wrong answer! Enter yes or no"exit 1
elif - Doing further Checks#!/bin/sh
echo "Is it morning? Please answer yes or no"read timeofday
if [ $timeofday = "yes" ]; thenecho "Good Morning"
elif [ $timeofday = "no" ]; thenecho "Good afternoon"
else
echo "Wrong answer! Enter yes or no"
exit 1
fi exit 0case
```

case *variable* in

pattern* [| *pattern*] ...) *statements;;pattern* [| *pattern*] ...) *statements;;

....

esac #!/bin/sh

echo "Is it morning? Enter yes or no";read timeofdaycase "$timeofday" in

yes | y | Yes | YES) echo "Good Morning";; n* | N*) echo "Good Afternoon";;

*) echo "Sorry, answer not recognized" echo "Please answer yes or no"

exit 1;;

esac

^C is 2 - SIGINT

FILE API FUNCTIONS

under this concept we discuss the functions or methods of regular file. On each function we discuss theusage, syntax, arguments, argument values and return types

Open():used to open a regular file.

Syntax: int open(const char *pathname, int oflag,/* mode_t mode*/);#include <sys/types.h>

#include <sys/stat.h>#include <fcntl.h>

int open(const char *pathname, int oflag,/* mode_t mode*/);

open -oflag

- O_RDONLY open for reading only
- O_WRONLY open for writing only
- O_RDWR open for reading and writing
- O_APPEND append on each write –not atomic when using NFS
- O_CREAT create file if it does not exist

•O_TRUNC truncate size to 0

•O_EXCL error if create and file exists

•O_SYNC Any writes on the resulting file descriptor will block the calling process until the datahas been physically written to the underlying hardware .

open -mode

•Specifies the permissions to use if a new file is created.

•This mode only applies to future accesses of the newly created file.

User: S_IRWXU, S_IRUSR, S_IWUSR, S_IXUSR**Group:** S_IRWXG,S_IRGRP,S_IWGRP,S_IXGRP **Other:** S_IRWXO, S_IROTH,S_IWOTH,S_IXOTH

•mode must be specified when O_CREAT is in the flags.

Identifying errors

•How can we tell if the call failed?

-the system call returns a negative number

•How can we tell what was the error?

-Using errno - a global variable set by the system call if an error has occurred.

-Defined in errno.h

-Use str error to get a string describing the problem

-Use p error to print a string describing the problem #include <errno.h>

```
int fd;

fd = open( FILE_NAME, O_RDONLY, 0644 );

if( fd < 0 ) {

printf( "Error opening file: %s\n", strerror( errno ) );return -1;

}
```

open –possible errno values

•EEXIST–O_CREAT and O_EXCL were specified and the file exists.

•ENAMETOOLONG -A component of a pathname exceeded {NAME_MAX} characters, or an entire path name exceeded {PATH_MAX} characters.

•ENOENT -O_CREAT is not set and the named file does not exist.

•ENOTDIR -A component of the path prefix is not a directory.

•EROFS -The named file resides on a read-only file system, and write access was requested.

•ENOSPC -O_CREAT is specified, the file does not exist, and there is no space left on the file systemcontaining the directory.

•EMFILE -The process has already reached its limit for open file descriptors.

create():used to create a regular file.

Syntax: int creat(const char *pathname, mode_t mode)#include <sys/types.h>

#include <sys/stat.h>#include <fcntl.h>

Int creat(const char *pathname, mode_tmode)

Equivalent to: open(pathname, O_WRONLY | O_CREAT | O_TRUNC, mode)

lseek:used to position the cursor at specified location.

Syntax: off_t lseek(int fd, off_t offset, int whence);#include <sys/types.h>

#include <unistd.h>

off_t lseek(int fd, off_t offset, int whence);

- fd

–The file descriptor.

–It must be an open file descriptor.

- offset

–Repositions the offset of the file descriptor fd to the argument offset according to the directivewhence.

• Return value

–The offset in the file after the seek

–If negative, errno is set.

lseek–whence

• SEEK_SET -The offset is set to offset bytes from the beginning of the file.

• SEEK_CUR -The offset is set to its current location plus offset bytes.

–currpos= lseek(fd, 0, SEEK_CUR)

• SEEK_END -The offset is set to the size of the file plus offset bytes.

–If we use SEEK_END and then write to the file, it extends the file size in kernel and the gap isfilled with zeros.

lseek: Examples

• Move to byte #16

–newpos= lseek(fd, 16, SEEK_SET);

• Move forward 4 bytes

–newpos= lseek(fd, 4, SEEK_CUR);

• Move to 8 bytes from the end

–newpos= lseek(fd, -8, SEEK_END);

• Move backward3 bytes

–lseek(fd, -3, SEEK_CUR);

lseek-errno

- lseek() will fail and the file pointer will remain unchanged if:

–EBADF - fd is not an open file descriptor.

–ESPIPE - fd is associated with a pipe, socket, or FIFO.

–EINVAL - Whence is not a proper value.

Read():used to read a block of data from regular file. Syntax: ssize_t read(int fd,void *buff,size_t nbytes)#include <unistd.h>

ssize_tread(int fd,void *buff,size_t nbytes)

• Attempts to read nbytes of data from the object referenced by the descriptor fd into the buffer pointed toby buff.

- If successful, the number of bytes actually read is returned.
- If we are at end-of-file, zero is returned.

• Otherwise, -1 is returned and the global variable errno is set to indicate the error.read -errno

• EBADF -fd is not a valid file descriptor or it is not open for reading.

• EIO -An I/O error occurred while reading from the file system.

• EINTR The call was interrupted by a signal before any data was read

• EAGAIN-The file was marked for non-blocking I/O, and no data was ready to be read.

write():used to write a block of data to the regular file. Syntax: ssize_t write(int fd, const void *buff, size_t nbytes) #include <unistd.h>

ssize_twrite(int fd, const void *buff, size_t nbytes)

• Attempts to write nbytes of data to the object referenced by the descriptor fd from the buffer pointed toby buff.

• Upon successful completion, the number of bytes actually written is returned.

–The number can be smaller than nbytes, even zero

• Otherwise -1 is returned and errno is set.

• "A successful return from write() does not make any guarantee that data has been committed to disk."write -errno

• EBADF - fd is not a valid descriptor or it is not open for writing.

• EPIPE -An attempt is made to write to a pipe that is not open for reading by any process.

• EFBIG -An attempt was made to write a file that exceeds the maximum file size.

• EINVAL -fdis attached to an object which is unsuitable for writing (such as keyboards).

• ENOSPC -There is no free space remaining on the file system containing the file.

• EDQUOT -The user's quota of disk blocks on the file system containing the file has been exhausted.

• EIO -An I/O error occurred while writing to the file system.

• EAGAIN -The file was marked for non-blocking I/O, and no data could be written immediately.

Example

```
#include <stdio.h> #include <sys/types.h>#include <sys/stat.h> #include <fcntl.h>

char buf1[] = "abcdefghij"; char buf2[] = "ABCDEFGHIJ";

int main(void) {int fd;

fd = creat("file.hole", S_IRUSR | S_IWUSR | IRGRP));if( fd< 0 ) {

perror("createrror");exit(1);

}

if( write(fd, buf1, 10) != 10 ) {perror("buf1 write error"); exit(1);

}

/* offset now = 10 */

if( lseek(fd, 40, SEEK_SET) == -1 )

{

perror("lseek error");exit(1);

}

/* offset now = 40 */ if(write(fd, buf2, 10) != 10)

{

perror("buf2 write error");exit(1);
```

```
}
/* offset now = 50 */exit(0);
}
```

Example –copying a file

```
#include <stdio.h> #include <errno.h> #include <stdlib.h>
#include <sys/types.h>#include <sys/stat.h> #include <fcntl.h>
enum {BUF_SIZE = 16};
int main(int argc, char* argv[])
{
int  fdread,  fdwrite; unsigned int total_bytes = 0;
ssize_t nbytes_read, nbytes_write;char buf[BUF_SIZE];
if (argc != 3) {
printf("Usage: %s source destination\n",argv[0]);
exit(1);
}
fdread = open(argv[1], O_RDONLY);if (fdread < 0) {
perror("Failed to open source file");exit(1);
}
fdwrite = creat(argv[2], S_IRWXU);if (fdwrite < 0) {
perror("Failed to open detination file");exit(1);
}
do {
nbytes_read = read(fdread, buf, BUF_SIZE);if (nbytes_read < 0) {
perror("Failed to read from file");exit(1);
}
nbytes_write = write(fdwrite, buf, nbytes_read);if (nbytes_write
< 0) {
perror("Failed to write to file");exit(1);
```

```
}
} while(nbytes_read > 0);close(fdread); close(fdwrite);
return 0;
}
```

dup2

```
#include <unistd.h>
Int dup2(int oldfd, int newfd);
```

- Duplicates an existing object descriptor and returns its value to the calling process.
- Causes the file descriptor newfd to refer to the same file as oldfd. The object referenced by the descriptordoes not distinguish between oldfd and newfd in any way.
- If newfd refers to an open file, it is closed first
- Now newfd and oldfd file position pointer and flags.

dup2 -errno

- EBADF - oldfd isn't an open file descriptor, or newfd is out of the allowed range for file descriptors.
- EMFILE -Too many descriptors are active.
- Note: If a separate pointer to the file is desired, a different object reference to the file must be obtained by issuing an additional open() call.

Dup2 -comments

- **dup2()** is most often used to redirect standard input or output.

~> ls | grep "my"

- dup2(fd, 0) - whenever the program tries toread from standard input, it will read from fd.
- dup2(fd, 1) -whenever the program tries to writeto standard output, it will write to fd.
- After arranging the redirections, the desired program is run using exec

fcntl: 1)used to change the access mode values2)used to duplicate the file descriptor

Syntax: int fcntl(int fd, int cmd, ...)#include <sys/types.h>

#include <unistd.h> #include <fcntl.h> intfcntl(intfd, intcmd, intarg)

- manipulate filedescriptors.
- cmd –the operation to perform
- arg –depends on the operation (not always required) fcntl-cmd
- F_DUPFD -Returns a new descriptor as follows:

–Lowest numbered available descriptor greater than or equal to arg.

–New descriptor refers to the same object as fd.

–New descriptor shares the same file offset.

–Same access mode (read, write or read/write).

- This is different fromdup2 which uses exactly the descriptor specified.
- F_GETFL -Returns the current file status flags as set by open().

–Access mode can be extracted from AND'ing the return value

- return_value & O_ACCMODE
- F_SETFL Set descriptor status flags to arg.

–Sets the file status flags associated with fd.

–Only O_APPEND, O_NONBLOCK and O_ASYNC may be set.25

fcntl–example 1

```
#include <stdio.h> #include <sys/types.h>#include <fcntl.h>
intmain( intargc, char *argv[] ){intaccmode, val;
if( argc!= 2 ) {
fprintf( stderr, "usage: %s <descriptor#>", argv[0] );exit(1);
}
```

```
val = fcntl(atoi(argv[1]), F_GETFL, 0);if (val< 0 ) {
perror( "fcntlerror for fd");exit( 1 );
}
accmode= val& O_ACCMODE;if( accmode == O_RDONLY ) printf( "read only");
else if(accmode == O_WRONLY )printf( "write only");
else if( accmode == O_RDWR )printf( "read write");
else {
fprintf( stderr, "unkown access mode");exit(1);
}
if( val & O_APPEND )
printf( ", append");
if( val & O_NONBLOCK)printf(", nonblocking");
if( val & O_SYNC )
printf(", synchronous writes");putchar( '\n');
exit(0);
}
```

fcntl-example 2 #include <stdio.h> #include <sys/types.h> #include <fcntl.h>

```
/* flags are file status flags to turn on */void set_fl( intfd, intflags ){
intval;
val = fcntl( fd, F_GETFL, 0 );if (val< 0 ) {
perror( "fcntlF_GETFL error");exit( 1 );
}
val |= flags;    /* turn on flags */ val = fcntl( fd, F_SETFL, val ) < 0if( val< 0 ) {
perror( "fcntlF_SETFL error");exit( 1 );
}
```

}

Links():used to create links between two files.

Syntax: int link(constchar *existingpath, constchar *newpath) (hard links and soft links)

#include <unistd.h>

Int link(constchar *existingpath, constchar *newpath) int symlink(constchar *actualpath, constchar *newpath); int unlink(constchar *pathname);

int remove(constchar *pathname);

int rename (const char *oldname, const char *newname);

link –make a hard link

int link(constchar *existingpath, const char *newpath)

• Makes a hard link to a file

• Atomically creates the specified directory entry (hard link) newpath with the attributes of the underlyingobject pointed at by existingpath.

- If the link is successful: the link count of the underlying object is incremented; newpath and existingpathshare equal access and rights to the underlying object.
- If existingpath is removed, the file newpath is not deleted and the link count of the underlying object isdecremented.

link() example

~>ls -ltotal 8

-rwx 1 jphb5804 Sep 25 15:44 mklink

-rw 1 jphb98 Sep 25 15:43 mklink.c

-r 1 jphb256 Sep 25 15:05 test

• Make a link in C:

link("test","new_name")

• Make a link in the shell:

~> ln test new_name

~>ls-ltotal 9

-rwx 1 jphb5804 Sep 25 15:44 mklink

-rw 1 jphb98 Sep 25 15:43 mklink.c

-r 2 jphb256 Sep 25 15:05 new name

-r 2 jphb256 Sep 25 15:05 test30

symlink –make a soft (symbolic) link

int symlink(constchar *actualpath, constchar *newpath);

- Makes a symboliclink to a file.
- To the normal user,a symbolic link behaves like a file,but the underlying mechanism is different.
- Creates a special type of file whose contents are the name of the target file
- The existence of the file is not affected by the existence of symbolic links to it.link() example

~> ls -ltotal 7

-rwx 1 jphb5816 Sep 29 14:04 mklink

-rw 1 jphb101 Sep 29 14:04 mklink.c

- **Make a link in C:**

symlink("test","new_name")

- Make a link in the shell:

~> ln –s test new_name

~> ls -ltotal 8

-rwx 1 jphb5816 Sep 29 14:04 mklink

-rw 1 jphb101 Sep 29 14:04 mklink.c

lrwxrwxrwx1 jphb4 Sep 29 14:04 new_name -> testDoesanyone see a problem here?32

Unlink(): used to remove the link. Syntax:int unlink(constchar *pathname);

•Removes the link (soft or hard) namedby pathname from its directory

- If a hard link, decrements the link count of the file which was referenced by the link.
- If that decrement reduces the link count of the file to zero, and no process has the file open, then allresources associated with the file are reclaimed.
- If one or more processes have the file open when the last link is removed, the link is removed, but theremoval of the file is delayed until all references to it have been closed.remove intremove(constchar

*pathname);

•Removes the fileor directory specified by path.

- If path specifies a directory, remove(path) is the equivalent of rmdir(path). Otherwise, it is the equivalentof unlink(path).

Rename intrename(constchar *oldname, const char *newname);

•Causes the link named oldname to be renamed as newname.

- If the file newname exists, it is first removed.

–The switch is done atomically

•Both oldname and newname must

–be bothdirectories or both non-directories

–resideon the same file system

–But they do not have to be on the same directory path

- If oldname is a symbolic link, the symbolic link is renamed, not the file or directory to which it points.

stat, fstat, lstat:used to retrieve the file attributes. Syntax: int stat(constchar *pathname, struct stat *buf)

intfstat(intfd, struct stat *buf)

Get information about a file #include <sys/types.h> #include <sys/stat.h>

intstat(constchar *pathname, struct stat *buf) intfstat(intfd, struct stat *buf) intlstat(constchar *pathname, struct stat *buf) **stat**

int stat(constchar *pathname, struct stat *buf)

• Obtains information about the file pointed to by pathname.

• Read, write or execute permission of the named file is not required, but all directories listed in the path name leading to the file must be searchable.

fstat

int fstat(intfd, struct stat *buf)

• Obtains the same information about an open file known by the file descriptor fd.Int lstat(constchar *pathname, struct stat *buf)

• like stat() except that ifthe named file is a symbolic link, lstat() returns information about the link, whilestat() returns information about the file the link references.

struct stat

Struct stat {

mode_tst_mode;/* file type and mode (type & permissions) */ ino_tst_ino; /* inode'snumber */

dev_tst_dev; /* device number (file system) */nlink_tst_nlink; /* number of links */ uid_tst_uid; /* user ID of owner */ gid_tst_gid; /* group ID */

off_tst_size; /* size in bytes */

time_tst_atime; /* last access */ time_tst_mtime; /* last modified */ time_tst_ctime; /* last file status change */long st_blksize; /* I/O block size */

long st_blocks; /* number of blocks allocated */

}

umask –user mask #include <sys/types.h> #include <sys/stat.h>

mode_tumask(mode_tcmask)

• Theumaskcommand permission bits to block when auser creates directories and files

• The bits inthe umaskare turned off from the modeargument to open.

• Example: Ifthe umaskvalue is octal 022, the results in new files being created with permissions 0666 is0666 &

~0022 = 0644 = rw-r--r--. 40

Chmod():used change the permissions.

Syntax: int chmod(constchar *pathname, mode_tmode)

#include <sys/types.h>#include <sys/stat.h>

intchmod(constchar *pathname, mode_tmode)

• Sets the file permission bits of the file specified by the pathname pathname to mode.

• User mustbe file ownerto change mode

Chown():used to change the owner of the file.

Syntax: intchown(constchar *pathname,uid_towner,gid_tgroup);

#include <sys/types.h>#include <unistd.h>

intchown(constchar *pathname,uid_towner,gid_tgroup);

• The owner ID and group ID of the file named by pathname is changed as specified by the argumentsowner and group.

• The owner of a file may change the group.

• Changing the owner is usually allowed only to the superuser.

FILE LOCKS:

File locking is a mechanism that restricts access to a computer file by only allowing one user or process access at any specific time. Systems implement locking to prevent the classic *interceding update* scenario the following example illustrates the *interceding update problem*:

1. Process A reads a customer record from a file containing account information, including the customer's account balance and phone number.

2. Process B now reads the same record from the same file so it has its own copy.
3. Process A changes the account balance in its copy of the customer record and writes the record back to the file.
4. Process B—which still has the original *stale* value for the account balance in its copy of the customer record—updates the customer's phone number and writes the customer record back to the file.
5. Process B has now written its stale account-balance value to the file, causing the changes madeby process A to be lost.

To over come the problem they introduced file locks.

1) Readlock: If a process applies read lock on data it prevents writing to the data by another processbut allows another process to read, also called as shared lock.

2) Writelock: if a process applies write lock it prevents another process from read and write. This alsocalled as exclusive lock.

There are two types of locking mechanisms: mandatory and advisory. Mandatory systems will actually prevent read()s and write()s to file.

Advisory Lock:

With an advisory lock system, processes can still read and write from a file while it's locked. Useless?Not quite, since there is a way for a process to check for the existence of a lock before a read or write.See, it's a kind of *cooperative* locking system. This is easily sufficient for almost all cases where file locking is necessary.

Mandatory Lock:

here are two types of (advisory!) locks: read locks and write locks (also referred to as shared locks and exclusive locks, respectively.) The way read locks work is that they don't interfere with other read locks. For instance, multiple processes can have a file locked for reading at the same. However, when a process has an write lock on a file, no other process can activate either a read or write lock until it is relinquished. One easy way to think of this is that there can be multiple readers simultaneously, but there can only be one writer at a time.

Setting a lock

The **fcntl()** function does just about everything on the planet, but we'll just use it for file locking. Setting the lock consists of filling out a struct flock (declared in *fcntl.h*) that describes the type of lock needed, **open()**ing the file with the matching mode, and calling **fcntl()** with the proper arguments, *comme* ça:

struct flock fl;int fd;

fl.l_type = F_WRLCK; /* F_RDLCK, F_WRLCK, F_UNLCK */ fl.l_whence = SEEK_SET; /* SEEK_SET, SEEK_CUR, SEEK_END */

fl.l_start= 0; /* Offset from l_whence */

fl.l_len = 0; /* length, 0 = to EOF */

fl.l_pid = getpid(); /* our PID */ fd = open("filename", O_WRONLY);

fcntl(fd, F_SETLKW, &fl); /* F_GETLK, F_SETLK, F_SETLKW */

What just happened? Let's start with the struct flock since the fields in it are used to *describe* the locking action taking place. Here are some field definitions:

l_type This is where you signify the type of lock you want to set. It's either F_RDLCK,F_WRLCK, or F_UNL-CK if you want to set a read lock, write lock, or clear thelock, respectively.

l_whence This field determines where the *l_start* field starts from (it's like an offset for the offset). It can be either SEEK_SET, SEEK_CUR, or SEEK_END, for beginning offile, current file position, or end of file.

l_start This is the starting offset in bytes of the lock, relative to *l_whence*.

l_len	This is the length of the lock region in bytes (which starts from *l_start* which isrelative to *l_whence*.
l_pid	The process ID of the process dealing with the lock. Use **getpid()** to gethis.

In our example, we told it make a lock of type F_WRLCK (a write lock), starting relative to SEEK_SET (the beginning of the file), offset 0, length 0 (a zero value means "lock to end-of-file), with the PID set to **getpid()**.

The next step is to **open()** the file, since **flock()** needs a file descriptor of the file that's being locked. Note that when you open the file, you need to open it in the same *mode* as you have specified in the lock, as shown in the table, below. If you open the file in the wrong mode for a given lock type, **fcntl()** will return

-1 and *errno* will be set to EBADF.

l_type	***mode***
F_RDLCK	O_RDONLY or O_RDWR
F_WRLCK	O_WRONLY or O_RDWR

Finally, the call to **fcntl()** actually sets, clears, or gets the lock. See, the second argument (the *cmd*) to **fcntl()** tells it what to do with the data passed to it in the struct flock. The following list summarizes whateach **fcntl()** *cmd* does:

F_SETLKW	This argument tells **fcntl()** to attempt to obtain the lock requested in the struct flock structure. If the lock cannot be obtained (since someone else has it locked already),**fcntl()** will wait (block) until the lock has cleared, then will set it itself. This is a very useful command. I use it all the time.
F_SETLK	This function is almost identical to F_SETLKW. The only difference is that this one will not wait if it cannot obtain a lock. It will return immediately with -1. This function can beused to clear a lock by setting the *l_type* field in the struct flock to F_UN-LCK.

F_GETLK — If you want to only check to see if there is a lock, but don't want to set one, you can use this command. It looks through all the file locks until it finds one that conflicts with the lock you specified in the struct flock. It then copies the conflicting lock's information into the struct and returns it to you. If it can't find a conflicting lock, **fcntl()** returns the struct as you passed it, except it sets the *l_type* field to F_UNLCK.

In our above example, we call **fcntl()** with F_SETLKW as the argument, so it blocks until it can set the lock, then sets it and continues.

1.2. Clearing a lock

Whew! After all the locking stuff up there, it's time for something easy: unlocking! Actually, this is a piece of cake in comparison. I'll just reuse that first example and add the code to unlock it at the end:

```
struct flock fl;

int fd;

fl.l_type   = F_WRLCK;  /* F_RDLCK, F_WRLCK, F_UNLCK
*/ fl.l_whence = SEEK_SET; /* SEEK_SET, SEEK_CUR, SEEK_END
*/

 fl.l_start  = 0;   /* Offset from l_whence         */
 fl.l_len = 0;      /* length, 0 = to EOF        */

fl.l_pid  =  getpid(); /* our PID
*/

fd = open("filename", O_WRONLY); /* get the file descriptor */
fcntl(fd, F_SETLKW, &fl);  /* set the lock, waiting if necessary */

.

.

.
```

```
fl.l_type  = F_UNLCK;  /* tell it to unlock the region */
fcntl(fd, F_SETLK, &fl); /* set the region to unlocked */
```

Now, I left the old locking code in there for high contrast, but you can tell that I just changed the *l_type*field to F_UNLCK (leaving the others completely unchanged!) and called **fcntl()** with F_SETLK as thecommand. Easy!

1.3. A demo program

Here, I will include a demo program, *lockdemo.c,* that waits for the user to hit return, then locks its own source, waits for another return, then unlocks it. By running this program in two (or more) windows, youcan see how programs interact while waiting for locks.

Basically, usage is this: if you run **lockdemo** with no command line arguments, it tries to grab a write lock (F_WRLCK) on its source (*lockdemo.c*). If you start it with any command line arguments at all, ittries to get a read lock (F_RDLCK) on it.

```
#include <stdio.h> #include <stdlib.h> #include <errno.h>
#include <fcntl.h> #include <unistd.h>

int main(int argc, char *argv[])

{

/* l_type  l_whence  l_start  l_len  l_pid */  struct flock fl =
{F_WRLCK, SEEK_SET,  0,  0,
0 };int fd;

fl.l_pid = getpid();if (argc > 1)

fl.l_type = F_RDLCK;

if ((fd = open("lockdemo.c", O_RDWR)) == -1) {perror("open");

exit(1);

}

printf("Press <RETURN> to try to get lock: ");getchar();

printf("Trying to get lock...");

if (fcntl(fd, F_SETLKW, &fl) == -1) {perror("fcntl");
```

```
exit(1);
}
printf("got lock\n");
printf("Press <RETURN> to release lock: ");getchar();
fl.l_type = F_UNLCK; /* set to unlock same region */if (fcntl(fd,
F_SETLK, &fl) == -1) {
perror("fcntl");
exit(1);
}
printf("Unlocked.\n");close(fd);
return 0;
}
```

Directory file API:

mkdir : This creates a new directory with no initial contents (apart from `.' and `..').

It returns -1 on failure

int mkdir(char *path, mode_t mode); path:specifies where to create and provide name

Example: mkdir(char "\home\student\credir",mode-t S_IRONLY)
rmdir: Removing a directory will fail unless the directory is empty.
int rmdir(char *path);

opendir: Open the directory specified by the path.DIR *opendir(char *path);

Closedir: Close the directory specified by directory file descriptor.

int closedir(DIR *dirp);

readdir: Read the directory specified by the directory file descriptor. If already create and open.struct dirent *readdir(DIR *dirp);

Objective Questions:

1. UNIX uses ls to list files in a directory. The corresponding command in MSenvironment is:

a. If b. listdir c. dir

2. A file with extension .txt

a. Is a text file created using vi editor b. Is a text file created using a notepad

c. Is a text file created using word

3. In the windows environment file extension identifies the application that created it. If we remove the file extension can we still open the file?

a. Yes b. No

4. Which of the following files in the current directory are identified by the regularexpression a?b*.

a. afile b. aab c. abb d. abc e. axbb f. abxy

5. For some file the access permissions are modified to 764. Which of the followinginterpretation are valid:

a. Every one can read, group can execute only and the owner can read andwrite.

b. Every one can read and write, but owner alone can execute.

c. Every one can read, group including owner can write, owner alone can execute

6. The file's properties in Windows environment include which amongst the following: Operating Systems/File Systems and Management Multiple Choice Questions

   a. File owners' name b. File size c. The date of last modification d. Date of file creation

   e. The folder where it is located

7. Which of the following information is contained in inode structure

   a. The file size b. The name of the owner of the file

   c. The access permissions for the file d. All the dates of modification since the file's creation

   e. The number of symbolic links for this file

8. File which are linked have as many inodes as are the links.

   a. True b. False

9. Which directory under the root contains the information on devices

   a. /usr/bin b. /usr/sbin c. /usr/peripherals/dev d. /etc/dev

10. A contiguous allocation is the best allocation policy. (True / False)

11. An indexed allocation policy affords faster information retrieval than the chainedallocation policy.

    a. True b. False

12. Absolute path names begin by identifying path from the root.

    a. True b. False

UNIT : III

Process Architecture and Control

Introduction

In a typical operating systems course, a process is dened to be a program in execution, or something similar to that. This is a true and accurate abstraction. A program such as the bash shell can have many, many instances running on a multi-user machine and each individual instance is a separate and distinct process, although each and every one of these is executing the exact same executable le. What this denition does not tell you is what a process is in concrete terms. It is like saying that a baseball game is an instance of the implementation of a set of rules created by Alexander Cartwright in 1845 by which two teams compete against each other on a playing eld. Neither denition gives you a mental picture of the thing being dened.

In this chapter, we focus on the concrete representation of a process in UNIX: how it is represented within the kernel, what kinds of resources it requires, how those resources are materialized and managed, what attributes it has, and what system calls are related to its control and management. As a rst step we will look at processes from the command level. Afterward, we will look at how UNIX systems arrange their address spaces and manage them in virtual and physical memory. Then we will look at how processes are created and how they communicate and synchronize with each other.

Examining Processes on the Command Line

The ps command is used for viewing one or more processes currently known to the operating system. I say "currently known" as opposed to "active" because the list may include processes that are technically not active, so called zombie processes. The set of options available for the ps command is very system dependent. There were dierent versions of it in BSD systems and in Version 7, and then there are options added by GNU. RedHat Linux systems support all of the historical options, and so there are many dierent ways to use this command in Linux. In Linux, users also have the option of running the top command. The top command is very much like ps, except that it displays the dynamic, real-time view of the state of the system. It is like the Windows task manager, and also like a terminal-based version of the Gnome System Monitor. Here we will describe the standard syntax rather than the BSD style or GNU syntax.

The ps command without options displays the list of processes of the login id that executes it, in a short form:

```
$ ps
  PID TTY          TIME CMD
  14244   pts/1    00:00:00 bash
  14572   pts/1    00:00:00 ps
```

You will notice that it always contains a line for the command itself because it itself has to run to do its job, obviously. The -f option causes it to display a full listing:

```
$ ps -f
UID     PID   PPID  C   STIME TTY    TIME CMD
sweiss  2508  2507  0   12:09 pts/8  00:00:00 -bash
sweiss  3132  2508  0   12:22 pts/8  00:00:00 ps -f
```

The UID column is the user login name. The PID column is the process id of the process. The PPID column is the parent process's process id. The C eld is rarely of interest; it gives processor utilization information. The STIME eld is the starting time in hours, minutes, and seconds. The TIME column is the cumulative execution time, which for short commands will be zero, since the smallest unit of measurement in this column is seconds. The CMD column is the command line being executed by this process; in this case there are two: bash and "ps -f". All command line arguments are displayed; if any are suppressed, the command will appear in square brackets:

```
root     3080    1       0 Jan29 ?        00:00:00 [lockd]
```

The TTY column is the controlling terminal attached to this process. Some processes have no terminal, in which case a "?" will appear.

The -e option displays all processes, which will be quite long. If I want to know which process is the parent of my bash process, I can use ps -ef and lter using grep:

```
$ ps -ef | grep 2507
sweiss  2507   2504   0       12:09 ?  00:00:00 sshd: sweiss@pts/8
sweiss  2508   2507   0       12:09    pts/8   00:00:00 -bash
sweiss  3207   2508   0       12:30    pts/8   00:00:00 grep 2507
```

From this output you see that I am connected via ssh on pseudo-terminal pts/8 and that the ssh daemon sshd is the parent of my bash process.

You can learn a lot about a system just by running ps. For example, on our Linux system, the rst few processes in the system are:

```
$ ps -ef | head -4
UID     PID     PPID    C STIME TTY  TIME CMD
root    1       0       0      Jan29 ? 00:00:01 init [5]
root    2       1       0      Jan29   ?       00:00:02 [migration/0]
root    3       1       0      Jan29   ?       00:00:00 [ksoftirqd/0]
```

whereas on our Solaris 9 server, the list is:

```
$ ps -ef | head -4
UID     PID PPID        C                STIME TTY       TIME CMD
root    0       0       0       Mar 13 ?         0:23     sched
root    1       0       0       Mar 13 ?         0:00     /etc/init -
root    2       0       0       Mar 13 ?         0:00     pageout
```

Notice that in Solaris, the (CPU) process scheduler itself is the very rst process in the system. It is absent in Linux. In all UNIX systems, the process with PID 1 is always init. In Solaris, the pageout process is responsible for writing pages to disk, and fsflush ushes system buers to disk.

The -u and -U options are useful for viewing all of your processes or those of others in a supplied user list. The list of users must be comma-separated, with no intervening spaces. For example:

```
$ ps -f -U sweiss,wgrzemsk
UID       PID   PPID  C STIME TTY    TIME CMD
sweiss    2507  2504  0 12:09 ?      00:00:00 sshd: sweiss@pts/8
sweiss    2508  2507  0 12:09 pts/8  00:00:00 -bash
wgrzemsk  2572  2570  0 12:10 ?      00:00:00 sshd: wgrzemsk@notty
wgrzemsk  2575  2573  0 12:10 ?      00:00:00 /bin/sh
```

While there are dozens of other options, I will only mention one more: the -o option. You can customize the output of the ps command to include any of the dozens of attributes available to be displayed using -o. The man page gives the general format for this. Some examples from the man page:

```
ps -eo pid,tid,class,rtprio,ni,pri,psr,pcpu,stat,wchan:14,comm ps axo
stat,euid,ruid,tty,tpgid,sess,pgrp,ppid,pid,pcpu,comm ps -
eopid,tt,user,fname,tmout,f,wchan
```

Note that there are no spaces in the list. In general never use spaces in any of the lists because the shell will then treat them as separate words rather than a single word to be passed to the ps command itself.

A related command is pgrep. If you need the process id of a command or program that is running, typing pgrep <executable name> will give you a list of processes running that program, one per line. For example

```
$ pgrep bash
2508 3502
3621
```

showing that three instances of bash are running, with pids 2508, 3502, and 3621.

Process Groups

UNIX systems allow processes to be placed into groups. There are several reasons for grouping processes. One is that a signal can be sent to an entire process group rather than a single process. For example, the shell arranges that all processes created in order to carry out the command line are in a single group, so that if the user needs to terminate that command, a single signal sent via Ctrl-C will kill all processes in the group. The alternative would require using the ps command to nd all processes that were created to carry out the command.

Every process has a process group-id (of type pid_t). There is a single process in each group that is considered to be the leader of the group. It can be identied easily, because it is the only process whose process group-id is the same as its process-id . You can view the process group-id of a process in the output of ps by using the -o option and specifying the format in either AIX format, such as

ps -o'%U %p %P %r %C %x %y %a'

or in standard format, as in

ps -ouser,pid,ppid,pgrp,%cpu,cputime,tty,args

If you run the command

$ cat | sort -u | wc

and then view the processes using one of the above ps commands, you will see the group formation:

$ psg -u sweiss | egrep 'TIME|cat|sort|wc'

USER PID PPID PGID %CPU TIME TTY COMMAND sweiss 17198 17076 17198 0.0 00:00:00 pts/2 cat sweiss 17199 17076 17198 0.0 00:00:00 pts/2 sort -u sweiss 17200 17076 17198 0.0 00:00:00 pts/2 wc

Notice that the cat command's process group-id (pgid) is the same as its process-id (pid) and that the three processes belong to the same group. If the full listing were displayed you would see that no other process is in this group.

Foreground and Background Processes

UNIX allows processes to run in the foreground or in the background. Processes invoked from a shell command line are foreground processes, unless they have been explicitly placed into the background by appending an ampersand '&' to the command line. There can be only one process group in the foreground at any time, because you cannot enter a new command until the currently running one terminates and the shell prompt returns. Foreground processes can read from and write to the terminal.

In contrast, there is no limit to the number of background processes, but in a POSIX-compliant system, they cannot read from or write to the terminal. If they try to do either, they will be stopped by the kernel (via SIGTTIN or SIGTTOU signals). The default action of a SIGTTOU signal is to stop the process, but many shells override the default action to allow background processes to write to the terminal.

Background and foreground processes use the terminal as their control terminal, but background processes do not receive all signals from that terminal. A Ctrl-C, for example, will not cause a SIGINT to be sent to background processes. However, a SIGHUP will be sent to all processes that use that terminal as their control terminal, including background processes. This is so that, if a terminal connection is broken, all processes can be notied of it and killed by default. If you want to start a background process and then logout from a session, you can use the nohup command to run it while ignoring SIGHUP signals, as in

```
$ nohup do_backup &
```

which will let do_backup run after the terminal is closed. In this case, the do_backup program must not read or write a terminal.

Sessions

Every process belongs to a session. More accurately, every process group belongs to a session, and by transitivity, each process belongs to a session. Every session has a unique session-id of type pid_t. The primary purpose of sessions is to organize processes around their controlling terminals. When a user logs on, the kernel creates a session, places all processes and process groups of that user into the session, and links the session to the terminal as its controlling terminal. Sessions usually consist of a single foreground process group and zero or more background process groups. Just as process groups have leaders, sessions have leaders. The session leader can be distinguished because its process-id is the same as the session-id.

Processes may secede from their sessions, and unlike countries, they can do this without causing wars. Any process other than a process group leader can form a new session and automatically be placed into a new group as well. The new session will have no control terminal. This is exactly how a daemon is created it detaches itself from the session into which it was born and goes o on its own. Later we will see how programs can do this.

You can add output to the ps command to see the session-id by adding the "sid" output format to the standard syntax, as in

```
$ ps -ouser,pid,ppid,pgrp,sid,%cpu,cputime,tty,args
```

The Memory Architecture of a Process

Although earlier chapters made allusions as to how a process is laid out in virtual memory, here the process layout is described in detail. In addition, we provide a program that displays enough information about the locations of its own symbols in virtual memory that one can infer its layout from them. In particular, the program will display the addresses of various local and global symbols in hexadecimal and decimal notation. These locations are clues to how the process is laid out in its logical address space.

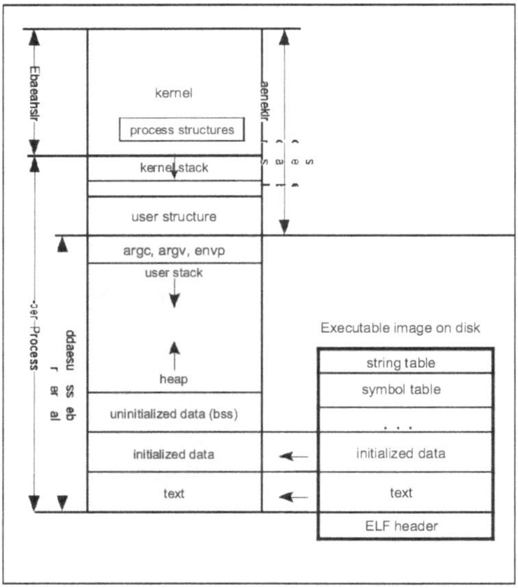

Figure 3.1: Typical layout of a process in virtual memory.

Overview

To start, we look at the big picture of what a process looks like in its logical address space. This picture should enable you to construct a mental image of the physical layout of a process in its own logical address space, how it looks in a le, and how that le relates to the virtual memory image of the process. Dierent UNIX systems use dierent layouts for processes, but for the most part, most modern systems adhere to a format known as the Executable and Linkable Format (ELF).

The resources needed by a process executing in user mode include the CPU state (general purpose registers, program status register, stack related registers, etc.), some environment information, and three memory segments called the text segment, the data segment, and the stack segment. The text segment contains the program code. The stack segment is reserved for the run-time stack of the user phase1 of the process, and the data segment is for program data. More will be said about these below.

The resources needed when the process is executing in kernel phase include the CPU state (same as above) as well as the resources needed for the kernel to provide the services for the process and schedule it appropriately. These resources include the parameters to the system call, the process's identity and properties, scheduling information, open le descriptors, and so on. This set of kernel resources is separated into two main structures, a process structure and a user structure, and a few minor ones. Together these structures constitute what is usually called the process image. The process structure and user structure are kept in kernel memory. The layout of the process address space is shown in Figure 3.1.

The process structure contains the information that must be memory-resident even when the process is swapped out, including the process privileges and rights, identiers associated with the process, its memory map, descriptors, pending events, and maximum and current resource utilization. The user structure contains information that is not needed in memory when the process is swapped out, including the process control block 2, accounting and statistics, and a few other pieces of information, and is therefore swapped out along with the rest of the process.

The Process Structure

The kernel maintains a process structure for every running process. This structure contains the infor-mation that the kernel needs to manage the process, such as various process descriptors (process-id, process group-id, session-id, and so on), a list of open les, a memory map and possibly other process attributes. In all of the versions of UNIX with which I am familiar, this structure is known as the process structure. In Linux, the terms process structure and task structure are used interchangeably, and the actual data structure that represents it is the task_struct . In some versions of UNIX, there is much less information in the process structure and more in the user structure.

Up until the introduction of threads, or so called light-weight processes (LWPs), the process structure was a very heavy structure lled with a large amount of information. One reason that the concept of a light-weight process was invented was to reduce the amount of information associated with each executable unit, so that they did not take up as much memory and so that creating new ones would be faster. The process structure was redesigned in 4.4BSD to support these threads by moving much of the information that had been in it into smaller structures that could be pointed to by the process structure. Each thread could share the information in the substructures by pointing to them, rather than keeping complete copies of them. This way, each thread could have its own unique identiers, such as a process-id, and also have access to the shared data, such as open les and memory maps.

The exact information present in any process structure will vary from one implementation to another, but all process structures minimally include

Process id

Parent process id (or pointer to parent's process structure)

Pointer to list of children of the process

Process priority for scheduling, statistics about CPU usage and last priority. Process state

Signal information (signals pending, signal mask, etc.)
Machine state
Timers

Usually the process structure is a very large object containing much more additional information. The task structure in Linux 2.6.x, for example, may contain over 150 dierent members. Typical substructures referenced in the process structure may include such things as the

Process's group id
User ids associated with the process
Memory map for the process (where all segments start, and so on) File descriptors
Accounting information
Other statistics that are reported such as page faults, etc.
Signal actions
Pointer to the user structure.

The substructure generally contains information that all threads would share, and the process structure itself contains thread-specic information.

The process structure is located in kernel memory. Dierent versions of UNIX store it in dierent ways. In BSD, the kernel maintains two lists of process structures called the zombie list and the allproc list. Zombies are processes that have terminated but that cannot be destroyed, the reasons for which will be made clear a little later in this chapter. Zombie processes have their structures on the zombie list. The allproc list contains those that are not zombies. In Linux 2.6.x, the process or task structures are kept in one, circular, doubly-linked list. In Solaris, the process structure is in a struct proc_t and the collection of these are maintained in a table.

The User Structure

The user structure contains much less information than the process structure. The user structure gets swapped in and out with the process; keeping it small reduces the swapping overhead. Historically, the most important purpose of the user structure was that it contained the per-process execution stack for the kernel.

The kernel phase begins, the kernel needs a stack for its function calls. Since the kernel might be interrupted in the middle of servicing a call, it must be able to switch from one process's service call to another. This implies that it needs a dierent stack for each process. The UNIX kernel designers carefully designed all functions in the kernel so that they are non-recursive and do not use large automatic variables. Furthermore, they can trace the possible sequences of call chains in the kernel so that they know exactly the largest possible stack size. Thus, unlike ordinary user programs, the kernel itself has a known upper bound on its required stack size. Because the stack size is known in advance, the kernel stack can be allocated in a xed size chunk of virtual address space. This is why, in Figure 3.1, you see that it is bounded above and below by xed boundaries. As you can see from the gure, the stack is at the high end of the address space, above the environment variables and program parameters. Its exact placement varies from one version of UNIX to another.

Depending on the version of UNIX, the user structure can contain various other pieces of information. In BSD, the memory maps are all in the process structure or its substructures. In Linux, the user structure, dened in struct user (in <user.h>) contains the memory maps. The memory maps generally include the starting and ending addresses of the text, data, and stack segments, the various base and limit registers for the rest of the address space, and so on1. The user structure usually contains the process control block. This contains the CPU state and virtual memory state (page table base registers and so on.)

The Text Segment

The text segment (also called the instruction segment) is the program's executable code. It is almost always a sharable, read-only segment, shared by all other processes executing the same program. The C compiler, by default, creates shared text segments. The advantage of using shared text segments is not so much that it conserves memory to do so, but that it reduces the overhead of swapping. When a process is active, its text segment resides in primary memory. There are various reasons why the process might be swapped to secondary storage. Often it is that it issues a wait for a slow event; in this case the system will swap it to secondary storage to make room in memory for other more productive processes. When it becomes active again, it is brought back into memory. Since a read-only text segment can never be modied, there is no reason to copy it to secondary storage when a process executing it is swapped out. Similarly, if a copy of a text segment already resides in primary memory, there is no reason to copy the text segment from secondary storage into primary memory. Thus, there is a savings in swapping overhead.

UNIX keeps track of the read-only text segment of each user process. It records the location of the segment in secondary storage and, if it is loaded, its primary memory address, and a count of the number of processes that are currently executing it. When a process rst executes the segment, the segment is loaded from secondary storage, the count is set to one, and a table entry is created. When a process terminates, the count is decremented. When the count reaches zero, the segment is freed and its primary and secondary memory are de-allocated.

The Stack Segment

The user stack segment serves as the run-time stack for the user phase of the process. That is, when the process makes calls to other parts of the user code, the calls are stacked in this segment. The stack segment provides storage for the automatic identiers and register variables, and serves its usual role of managing the linkage of subroutines called by the user process. The stack is always upside-down in UNIX, meaning that pushes cause the top to become a smaller memory address. If the stack ever meets the top of the heap, it causes an exception. In a 32-bit architecture, a user process is typically allocated a virtual address space of 4 Gbytes. If the stack meets the heap, the process exceeds its virtual memory allotment and it is time to port the application to a 64-bit machine!

The Data Segment

The data segment is memory allocated for the program's initialized and uninitialized data. The initialized data is separated from the uninitialized data, which is stored in a section called the bss, which is an acronym for Block Started by Symbol, an old FORTRAN machine instruction. Initialized data are items such as named constants and initialized static variables. They come from the symbol table. Uninitialized data has no starting value. The system only needs to reserve the space for them, which it does by setting the address of the top of the data segment. The data segment grows or shrinks by explicit memory requests to shift its boundary. The system call to shift the boundary is the brk() call; sbrk() is a C wrapper function around brk().The data segment always grows toward the high end of memory, i.e., the brk() call increases the boundary to increase memory. Most programmers use the C library functions malloc() and calloc() to allocate more memory instead of the low-level brk() primitives. These C routines call brk() to adjust the size of the data segment.

Programmers often refer to the part of memory that is used by malloc() and calloc(), the "heap". People usually think of the heap as the part of memory that is not yet allocated, lying between the top of the stack and the end of the bss.

From File to Memory

How is an executable program le arranged and how does it get loaded? When you run the gcc compiler and do not specically name the executable le, as in

```
$ gcc myprog.c
```

the compiler (actually the linker) creates a le named a.out. The name a.out is short for "assembler output" and was not just the name of the output le, but was also name of the format of all binary executable les on UNIX systems for many years. The a.out le format could be read in the a.out man page as well as in the <a.out> header le.

In the mid 1990's, a more portable and extensible format known as Executable and Linking Format (ELF) was created by the UNIX System Laboratories as part of the Application Binary Interface (ABI). It was later revised by the Tool Interface Standards (TIS) Committee, an industry con-sortium that included most major companies (Intel, IBM, Microsoft, and so on). While compilers continue to create les named a.out, they are ELF les on most modern machines.

The ELF specication denes exactly how an executable le is organized, and is general enough to encompass three dierent types of les:

Relocatable les holding code and data suitable for linking with other object les, to create an executable or a shared object le (.o les),

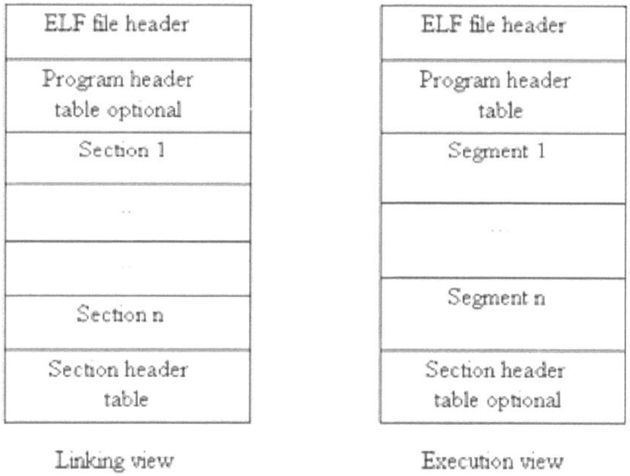

Figure : Linking and execution views of an ELF le.

Executable les holding a program suitable for execution, and Shared object les that can be used by the dynamic linker to create a process image (.so les).

An ELF le begins with a structure called the ELF header. This header is essentially a road map to the rest of the le. It contains identication information and the addresses and sizes of the rest of the components of the le. An ELF le is characterized by the fact that it contains two dierent, parallel, yet overlapping views: the linking view and the execution view as shown in Figure 7.2.

The linking view is the view of the le needed by the link editor in order to link and relocate components in the le. Information within it is organized into sections, which contain such things as the instructions, data, symbol table, string table, and relocation information. A section header table serves as a table of contents for the sections and is generally located at one end of the le.

The execution view, in contrast, is the view of the le needed in order to execute it. It organizes its information in segments.

Segments correspond conceptually to virtual memory segments; when the executable is loaded into memory, ELF segments are mapped to virtual memory segments. Thus, for example, for an executable program, there is a text segment containing instructions, an uninitialized data segment, and an initialized data segment, as well as several others. A program header table serves as a table of contents for the segments and usually follows the ELF header.

Neither the sections nor the segments in the le have to be in any particular order because the tables dene their positions. The two separate views are overlapped in the le, as shown in Figure. The gure also shows that segments can consist of multiple sections.

The symbol table is a table that the compiler creates to map symbolic names to logical addresses and store the attributes of these symbols. The compiler uses the table to construct the code, but in addition, it is the symbol table that makes it possible for a debugger to associate memory addresses with the names of variables. When the debugger runs, the symbol table is loaded into memory with the program. The string table is a table containing all of the strings used in the program. The strings command is a handy command to know about it displays a list of all of the strings in a binary le. The strings command works because the string table is part of the le and the command simple has examine it.

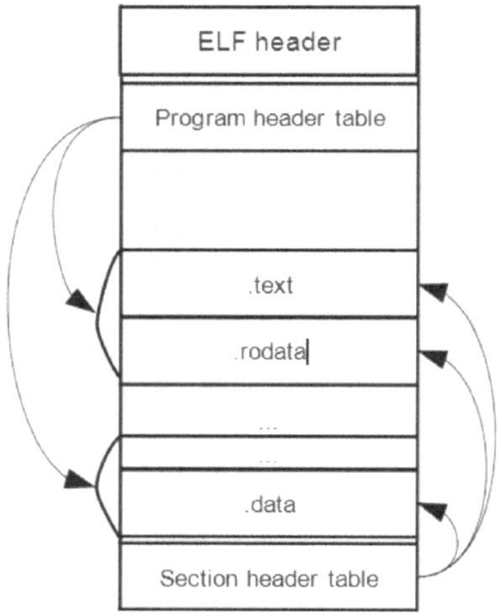

Figure: Overlapped views of the ELF le.

The readelf command can be used to examine an executable le.

$ readelf [options] elffile ...

The information displayed depends upon the options provided on the command line. With -a, all information is provided. The -h option displays the contents of the ELF header (in human readable form of course):

```
$   readelf -h /bin/bash ELF Header:
Magic: 7f 45 4c 46 01 01 01 00 00 00 00 00 00 00 00 00
Class: ELF32
Data: 2's complement, little endian
Version: 1 (current)
OS/ABI: UNIX - System V
ABI Version: 0
Type: EXEC (Executable file)
Machine: Intel 80386
Version: 0x1
Entry point address: 0x805c7b0
Start of program headers: 52 (bytes into file)
Start of section headers: 733864 (bytes into file)
Flags: 0x0
Size of this header: 52 (bytes)
Size of program headers: 32 (bytes)
Number of program headers: 8
Size of section headers: 40 (bytes)
Number of section headers: 32
Section header string table index: 31
```

You can see from this output that the ELF header has information about the executable's format: 32-bit, 2's complement, for Intel 80386 on UNIX System V. It also has the location of the program header (byte 52 into the le) and the section header table (733864 bytes into the le).

The file command uses the ELF header to provide its output:

```
$ file /bin/bash
/bin/bash: ELF 32-bit LSB executable, Intel 80386, version 1 (SYSV), dynamically linked (uses shared libs), for GNU/Linux 2.6.9, stripped
```

You can experiment with some of the executables to see how much you can learn about the structure of executable programs.

Notice in the output above that the entry point address of the executable is 0x805c7b0. This is the address of the rst executable instruction in the bash executable. The starting virtual address is not 0. In modern UNIX systems, the starting address is always after 0x8048000. The addresses below that are reserved for the system to use. One reason for this is that the debugger will run in the lower addresses when the program runs under the debugger.

Figure shows the virtual addresses of a hypothetical executable, taken from a version of the ELF standard. Notice in this example that the text segment is not at the start of the virtual address space, and that each segment is padded as needed so that it aligns on 0x1000 (4096) byte boundaries, because page sizes are 4096 bytes. Notice too that the data segment and uninitialized data segment follow the text segment.

A Program to Display The Virtual Memory Boundaries

In this section we will explore the virtual address space of a process with the aid of a program that I found in a book on interprocess communication [1] and subsequently modied. It displays the boundaries of the dierent components of the virtual address space of its executable image. The program declares the following types of memory objects:

Global, initialized pointer variable: cptr
Global uninitialized string: buffer1
Automatic variable in main program: i
Parameters to main program: argc, argv, envp
Static uninitialized local in main program: diff Main function main()
Non-main function showit()

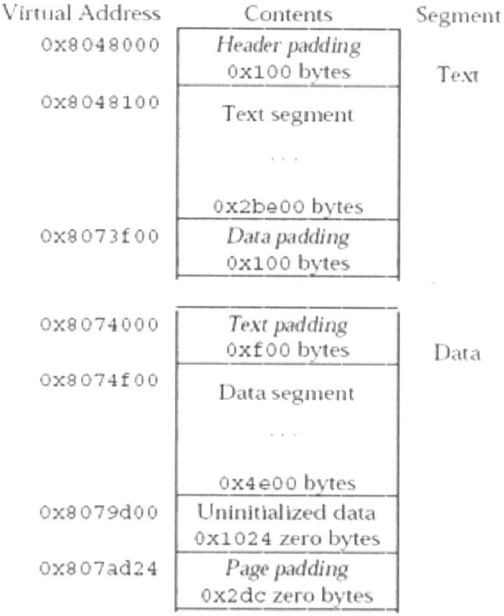

Figure : Example of ELF process image segments.

Automatic pointer variable, dynamically allocated: buffer2

Automatic variable in non-main function: num

It displays the locations of each of these objects as hexadecimal and decimal virtual addresses. The locations of these objects will lie within the ranges that are specied by the text, data, and stack segment positions described earlier. For example, the location of the uninitialized global bufferl and the initialized global *cptr will be between the rst and last addresses of the data segment, one in the bss and the other, not. The remaining variables each pinpoint the location of a particular segment: "Hello World ", the pointer variable cptr itself, the symbols main() and showit(), which are globals, and the local variables num and buffer2 in showit(). The location of a local variable, which is supposed to be on the run-time stack, will show where that stack is in virtual memory.

Three external integer variables etext, edata, and end are dened in global scope in the C library and may be accessed by any C program3. They are boundaries of three specic segments:

etext The address of etext is the rst location after the program text.

edata The address of edata is the rst location after the initialized data region.

end The address of end is the rst location after the uninitialized data region.

The value of etext is an upper bound on the size of the executable image. The initialized and uninitialized data regions are the regions where constants and globals are stored. Uninitialized data are globals whereas initialized data are constants. These symbols are incorporated into the program, displayed in Listing 3.1.

Listing 3.1: displayvm.c

```
#include<stdio.h>
#include<stdlib.h>
#include<string.h>
#include<sys/types.h>
#include<unistd.h>

/                                                            /
/                   Global   Constants                       /
//
#define SHW_ADR( ID , I ) \
printf("     %s \t  is  at addr :%8X\ t%20u\n " , \
       ID, (unsigned  int)(& I ), (unsigned int )(& I ) )

/ These  are system  variables, defined in the unistd.h header file   /
extern int   etext, edata ,  end ;

char     cptr=  "Hello World\n";    /  cptr  is  an initialized global   /
char     buffer1[40];               / uninitialized    global            /

void showit(char  );          / Function prototype     has  no storage   /
int main (int  argc,  char argv[],   char  envp[])
{
    int   i=0;              / on stack /
    static int    diff;     / global   in uninitialized  data  segment /
    strcpy(buffer1, "    Layout of virtual  memory\n " );
    write(1,buffer1,strlen(buffer1)+1);

    printf("Adr etext: %8X\t Adr  edata: %8X  \t Adr  end: %8X \n\n ",
           (unsigned  int)&etext,(unsigned   int) &edata , (unsigned int) &end );

    printf("     ID \t          HEX_ADDR\ t          DECIMAL_ADDR\n " ) ;
   SHW_ADR(" main ",  main );
   SHW_ADR("showit", showit);
   SHW_ADR( "etext",etext);
    diff =(int)  &showit  (int)&main;
    printf("          showit is%d bytes above main\n ",  diff);
   SHW_ADR( "cptr", cptr);
    diff=(int)&cptr     (int)&showit;
    printf("          cptr  is%d  bytes above  showit\n",  diff);
   SHW_ADR("buffer1",buffer1);
   SHW_ADR("diff",  diff);
   SHW_ADR("edata", edata);
   SHW_ADR(" end ", end );
   SHW_ADR("argc",  argc);
   SHW_ADR(" argv ", argv );
   SHW_ADR(" envp ", envp );
   SHW_ADR("i",  i);
```

```
    showit(cptr);
    return 0;
}
/                                                              /

void showit(char p)
{
    char  buffer2;
   SHW_ADR("buffer2",buffer2);
    if (  (buffer2=(char ) malloc((unsigned)(strlen(p)+1))) != NULL){
        strcpy(buffer2, p);
        printf("%s", buffer2);
        free(buffer2);
    }
    else {
        printf("Allocation error.\n");
        exit(1);
    }
}
```

When you run this program, you should get output similar to the following. The last column is the decimal value of the location. Notice that i in main() and buffer2 in showit() are high addresses. They are in the stack. Notice that envp, which is a pointer to an array of strings, is above the stack, as Figure 7.1 depicts. Notice that the addresses of these descend, because the stack grows downward. Notice too that diff in main() and buffer1 lie between edata and end, showing that they are in the uninitialized data region, but that cptr is between etext and edata because it is initialized. You should see that buffer1 is the last variable in the bss, since it is 40 bytes long and its address is 40 bytes from the end. Also, notice that the address of etext is the same as the value of etext. This is because etext is not a real variable. It is just a mnemonic name for the actual location, as determined by the linker.

```
      Layout of virtual memory
Adr etext:  8048958          Adr edata:  8049C90            Adr end:  8049CE8

        ID                       HEX_ADDR                 DECIMAL_ADDR
        main          is at addr:  8048544                   134513988
        showit        is at addr:  8048806                   134514694
        etext         is at addr:  8048958                   134515032
            showitis 706 bytes above main
        cptr          is at addr:  8049C8C                   134519948
            cptr is 5254 bytes above showit
        buffer1       is at addr:  8049CC0                   134520000
        diff          is at addr:  8049CA8                   134519976
        edata         is at addr:  8049C90                   134519952
        end           is at addr:  8049CE8                   134520040
        argc          is at addr: BF9ECDC0                  3214855616
        argv          is at addr: BF9ECDC4                  3214855620
        envp          is at addr: BF9ECDC8                  3214855624
        i             is at addr: BF9ECD9C                  3214855580
        buffer2       is at addr: BF9ECD6C                  3214855532
Hello World
```

Notice that the main program starts at virtual address 0x08048544. The start of the virtual address space is at 0x08048000 . The dierence is 1348 bytes. If you look at the output of the readelf-a command, you will see that there are several sections of code that reside in this small pocket before the main program. There you will nd the code that must be run before main() starts (what I like to call the "glue" routine), the code that runs when the program nishes, and special code for dynamic linking and relocation. When a program starts, before the operating system transfers control to the program, there are initializations and possible linking and relocation; the code there serves that purpose.

Creating New Processes Using fork

Now that you have a better idea of what processes actually are, we can start exploring their world.

We will begin with process creation, since that is, after all, the beginning of all things.

Once the operating system has bootstrapped itself, the only way for any process to be created is via the fork() system call4. All processes are created with fork():

```
#include <sys/types.h>
#include <unistd.h>
pid_t fork(void);
```

The fork() call is a hard one to accept at rst; you probably have never seen a function quite like it before. It is very appropriately named, because the statement

```
pid_t processid = fork();
```

causes the kernel to create a new process that is almost an exact copy of the calling process, such that after the call, there are two processes, each continuing its execution at the point immediately after the call in the executing program! So before the instruction is executed, there is a single process about to execute the instruction, and by the time it has returned, there are two. There has been a fork in the stream of instructions, just like a fork in a road. It is almost like process mitosis.

The process that calls fork() is called the parent process and the new one is the child process. They are distinguished by the values returned by fork(). Each process is able to identify itself by the return value. The value returned to the parent is the process-id of the newly created child process, which is never 0, whereas the value returned to the child is 0. Therefore, each process simply needs to test whether the return value is 0 to know who it is. The child process is given an independent copy of the memory image of the parent and the same set of resources. It is essentially a clone of the parent and is almost identical in every respect5. Unlike cellular mitosis though, it is not symmetric.

The typical use of the fork() call follows the pattern

processid = fork();
// child's code here
else
// parent's code here

The true branch of the if-statement is code executed by the child and not by the parent. The false branch is just the opposite. This may seem like a useless way to create processes, since they always have to share the same code as the parent, so they do nothing dierent. The fork() call is valuable when used with exec() and wait(), as will be shown shortly.

Our rst example, forkdemo1.c , in Listing 3.2 demonstrates a bit about how fork() works.

Listing 3.2: forkdemo1.c

```
#include<unistd.h>
#include<stdio.h>
#include<stdlib.h>

int global=10;

int main(int argc,  char argv[])
{
    int    local=0;
    pid_t  pid;

    printf("Parent process:(Before fork())");
    printf(" local=%d, global=%d\n", local, global);

    if ((pid=fork())== 1) {
        perror("fork");
        exit(1);
    }
```

```
    else if ( 0==pid ) {
        / child executesthis branch /
        printf("After the fork in the child:");
        local++;
        global++;
        printf(" local = %d, global = %d\n", local, global);
    }
    else {
        / parentexecutes this branch /
        sleep(2); / sleep long enough forchild's output to appear /
    }

    / both processes execute this print statement /
    printf("pid=%d, local=%d, global=%d\n",
            getpid(), local, global);

    return0;
}
```

In this program, the return value from fork() is stored in pid, which is tested in the if-statement. fork() returns -1 on failure, and one should always check whether it failed or not.

The child's code is the next branch, in which 0 == pid is true. The child increments the values of two variables, one, global, declared globally and one, local, locally in main(). It then prints out their values, jumps over the parent code and executes the printf(), obtaining its process-id using the getpid() system call and displaying the values of local and global again.

In the meanwhile, the parent sleeps for two seconds, enough time so that the child's output will appear rst. It then executes the printf(), obtaining its process-id using the getpid() system call and displaying the values of local and global. The sleep() prevents intermingling of the output, which can happen because the child shares the terminal with the parent.

Bearing in mind that the child is given a copy of the memory image of the parent, what should the output of the parent be? Hopefully you said that local = 0 and global = 10. This is because the child changed their values in the copy of the memory image, not in a shared memory. The point of this program is simply to demonstrate this important fact.

Let us make sure we understand fork() arithmetic before continuing. Before running the program forkdemo2.c shown in Listing 3.3, predict the number of lines of output. Do not redirect the output or pipe it. This will be explained afterward.

Listing 3.3: forkdemo2.c

```
#include<unistd.h>
#include<stdio.h>
#include<stdlib.h>

int main( int argc, char argv[] )
{
    int i;
    printf("About to create many processes...\n");
    for( i=0;  i<N;  i++)
        if ( 1==fork() )
            exit(1);

    printf("Process id = %d\n", getpid());
    fflush(stdout);        / force output before shell prompt    /
    sleep(1);              / give time to the shell to diplay prompt    /
    return 0;
```

Did you correctly predict?

Each time the bottom of the loop is reached, the number of processes in existence is twice what it was before the loop was entered, because each existing process executes the fork() call, making a copy of itself. If N were 1, the loop would execute once and there would be 2 processes, each

printing their process ids to the screen. If N were 2, the loop would execute a second time, and the 2 processes would make 2 more, 4 in total. In general, there will be 2N processes when the loop

nishes, and that many lines of output on the screen, together with the rst line,

About to create many processes...

for a total, when N = 4, of 17 lines. Now try redirecting the output of the program to wc, to make it easier to count how many lines are there.

```
$ forkdemo2 | wc
32      144     832
$
```

Why are there 32 lines and not 17? Try instead redirecting the output to a le and looking at it there:

```
$   forkdemo2 > temp; wc temp 24
    108 624 temp
$
```

If you look at the le temp, you will see something like

About to create many processes...
Process id = 6708
About to create many processes...
Process id = 6707
About to create many processes...
Process id = 6717
About to create many processes...
Process id = 6709
About to create many processes...
Process id = 6716
...

and there will be fewer than 16 lines with process ids. So what is going on here? Why is that line replicated for each process and why are there fewer than 16 lines stating the process ids?

Remember this important fact about fork(): when a process is created by a call to fork(), it is an almost exact duplicate of the original process. In particular it gets copies of all open le descriptors and naturally, all of the process's user space memory image.

What this means is that when a child process is created, its standard output descriptor points to the same open le structure as the parent and all other processes forked by the parent, and therefore the children and parent share the le position pointer.

Operations such as printf() are part of the C I/O library and act on objects of type FILE, which are called streams. The C I/O Library uses stream buering for all operations that act of FILE streams. (We noted this in Chapter 5.) There are three dierent kinds of buering strategies:

Unbuered streams : Characters written to or read from an unbuered stream are trans-mitted individually to or from the le as soon as possible.
Line buered streams : Characters written to a line buered stream are transmitted to the le in blocks when a newline character is found.
Fully buered streams : Characters written to or read from a fully buered stream are transmitted to or from the le in blocks of arbitrary size.

By default, when a stream is opened, it is fully buered, except for streams connected to terminal devices, which are line buered.

The buers created by the C I/O Library are in the process's own address space, not the kernel's address space. (When your program calls a function such as printf(), the library is linked into that program; all memory that it uses is in its virtual memory.) This means that when fork() is called, the child gets a copy of the parent's library buers, and all children get copies of these buers. They are not shared ; they are duplicated.

The C I/O library ushes all output buers

When the process tries to do output and the output buer is full.

When the stream is closed.

When the process terminates by calling exit().

When a newline is written, if the stream is line buered.

Whenever an input operation on any stream actually reads data from its le.

When fflush() is called on that buer.

As a corollary to the preceding statement, until the buer has been ushed, it contains all characters that were written to it since the last time it was ushed.

No C I/O Library function is atomic. It is entirely possible that output can be intermingled or even lost if the timing of calls by separate processes sharing a le position pointer leads to this.

Now we put these facts together. The forkdemo2 program begins with the instruction

```
printf("About to create many processes...\n");
```

If output has not been redirected, then stdout is pointed to a terminal device and it is line buered. The string "About to create many processes...\n" is written to the terminal and removed from the buer. When the process forks the children, they get empty buers and write their individual messages to the terminal. Unless by poor timing a line is written over by another process, each process will produce exactly one line of output. It is quite possible that this will happen if there are a large enough number, N, of processes, as the probability of simultaneous writes increases rapidly towards 1.0 as N increases.

Let us do a bit of mathematical modeling. The printf() instruction

printf("Process id = %d\n", getpid());

writes to standard output. If the fraction of time that each process spends in the portion of the printf() function in which a race condition might occur is p, then there is a probability of 1 (1 p) N that at least two processes are in that portion of code at the same time.

If, for example, p = 0:05 and N = 16, then the probability of a race (and hence lost output) is 1 0:9516 u 0:56.

If N = 32, it is 0:81 and when N = 64, it is 0:96. So you see that as the number of processes increases, it becomes almost inevitable that lines will be lost, regardless of whether they are written.

UNIT : IV

Processes Concepts

A process is more than just a program. Especially in a multi-user, multi-tasking operating systemsuch as Linux there is much more to consider. Each program has a set of data that it uses to do what it needs. Often, this data is not part of the program. For example, if you are using a text editor, the file you are editing is not part of the program on disk, but is part of the process in memory. If someone else were to be using the same editor, both of you would be using the same program. However, each of you would have a different process in memory. See the figure below to see how this looks graphically.

Kernel support for Process:

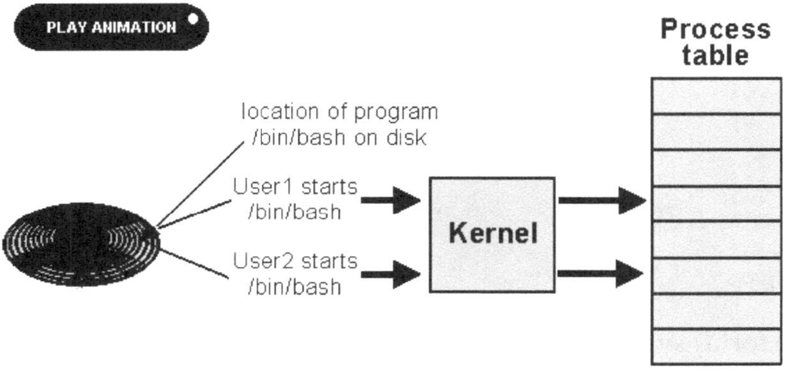

Image - Reading programs from the hard disk to create processes. (**interactive**)

Under Linux many different users can be on the system at the same time. In other words, they have processes that are in memory all at the same time. The system needs to keep track of what user is running what process, which terminal the process is running on, and what other resources the process has (such as open files). All of this is part of the process.

With the exception of the init process (PID 1) every process is the child of another process.

Another example we see in the next figure. When you login, you normally have a single process, which is your login shell(bash). If you start the X Windowing System, your shell starts another process, xinit. At this point, both your shell and xinit are running, but the shell is waiting for xinitto complete. Once X starts, you may want a terminal in which you can enter commands, so you start xterm.

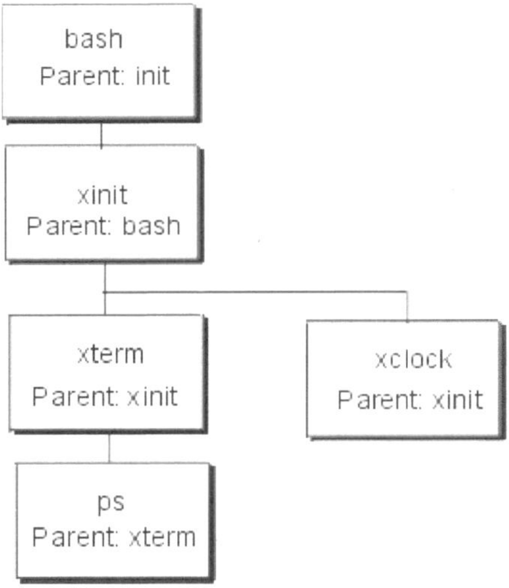

Process API

Fork():

The fork() system call will spawn a new child process which is an identical process to the parent except that has a new system process ID. The process is copied in memory from the parent and a new process structure is assigned by the kernel. The return value of the function is which discriminates the two threads of execution. A zero is returned by the fork function in the child's process.

exit() vs _exit():

The C library function exit() calls the kernel system call _exit() internally. The kernel system call _exit() will cause the kernel to close descriptors, free memory, and perform the kernel terminating process clean- up. The C library function exit() call will flush I/O buffers and perform aditional clean-up before calling

_exit() internally. The function exit(*status*) causes the executable to return "status" as the return code for main(). When exit(*status*) is called by a child process, it allows the parent process to examine the terminating status of the child (if it terminates first). Without this call (or a call from main() to return()) and specifying the status argument, the process will not return a value.

#include <stdlib.h>	#include <unistd.h>
void exit(int *status*);	void _exit(int *status*);

vfork():

The Vfork() function is the same as fork() except that it does not make a copy of the address space. The memory is shared reducing the overhead of spawning a new process with a unique copy of all the memory. This is typically used when using fork() to exec() a process and terminate. The vfork() function also executes the child process first and resumes the parent process when the child terminates.

wait(): Blocks calling process until the child process terminates. If child process has already teminated, the wait() call returns immediately. if the calling process has multiple child processes, the function returns when one returns.

waitpid(): Options available to block calling process for a particular child process not the firstone.

Kill():

This is the real reason to set up a process group. One may kill all the processes in the process group without having to keep track of how many processes have been forked and all of their process id's.

execl() and execlp():

The function call "execl()" initiates a new program in the same environment in which it is operating. An executable (with fully qualified path. i.e. /bin/ls) and arguments are passed to the function. Note that "arg0" is the command/file name to execute.

```
int execl(const char *path, const char *arg0,
const char *arg1, const char *arg2, ... const char
*argn, (char *) 0);
```

Where all function arguments are null terminated strings. The list of arguments is terminated byNULL.

The routine execlp() will perform the same purpose except that it will use environment variable PATH to determine which executable to process. Thus a fully qualified path name would not have to be used. The first argument to the function could instead be "ls". The function execlp() can also take the fully qualified name as it also resolves explicitly.

execv() and execvp():

This is the same as execl() except that the arguments are passed as null terminated array of pointers to char. The first element "argv[0]" is the command name.

```
int execv(const char *path, char *const argv[]);
```

The routine execvp() will perform the same purpose except that it will use environment variable PATH to determine which executable to process. Thus a fully qualified path name would not have to be used.The first argument to the function could instead be "ls". The function execvp() can also take the fully qualified name as it also resolves explicitly.

execve():

The function call "execve()" executes a process in an environment which it assigns.

Set the environment variables:

```
char *env[] = { "USER=user1", "PATH=/usr/bin:/bin:/
opt/bin", (char *) 0 };
```

Zombie Process

On Linux operating systems, a zombie process or defunct process is a process that has completed execution but still has an entry in the process table, allowing the process that started it to read its exit status. In the term's colorful metaphor, the child process has died but has not yet been reaped.

When a process ends, all of the memory and resources associated with it are deallocated so they can be used by other processes. However, the process's entry in the process table remains. The parent is sent a SIGCHLD signal indicating that a child has died; the handler for this signal will typically execute the waitsystem call, which reads the exit status and removes the zombie. The zombie's process ID and entry in theprocess table can then be reused. However, if a parent ignores the SIGCHLD, the zombie will be left in the process table. In some situations this may be desirable, for example if the parent creates another child process it ensures that it will not be allocated the same process ID.

A zombie process is not the same as an orphan process. Orphan processes don't become zombie processes; instead, they are adopted by init (process ID 1), which waits on its children.

The term zombie process derives from the common definition of zombie�an undead person.Zombies can be identified in the output from the Unix PS command by the presence of a "Z" in the STAT

column. Zombies that exist for more than a short period of time typically indicate a bug in the parent program. As with other leaks, the presence of a few zombies isn't worrisome in itself, but may indicate a problem that would grow serious under heavier loads.

To remove zombies from a system, the SIGCHLD signal can be sent to the parent manually, using the killcommand. If the parent process still refuses to reap the zombie, the next step would be to remove the parent process. When a process loses its parent, init becomes its new parent. Init periodically executes the wait system call to reap any zombies with init as parent.

Orphan Process

An orphan process is a computer process whose parent process has finished or terminated.

A process can become orphaned during remote invocation when the client process crashes after making arequest of the server.

Orphans waste server resources and can potentially leave a server in trouble. However there are several solutions to the orphan process problem:

1. Extermination is the most commonly used technique; in this case the orphan process is killed.
2. Reincarnation is a technique in which machines periodically try to locate the parents of any remote computations; at which point orphaned processes are killed.
3. Expiration is a technique where each process is allotted a certain amount of time to finish before being killed. If need be a process may "ask" for more time to finish before the allotted time expires.

A process can also be *orphaned* running on the same machine as its parent process. In a UNIX-like operating system any orphaned process will be immediately adopted by the special "init" system process.This operation is called re-parenting and occurs automatically. Even though technically the process has the "init" process as its parent, it is still called an orphan process since the process which originally created it no longer exists.

Inter Process Communication

Objective: how more than one process communicates with other processes sand calling functions, kernelsupport

IPC:Message Queues:<sys/msg.h>

The basic idea of a *message queue* is a simple one.

Two (or more) processes can exchange information via access to a common system message queue. The *sending* process places via some (OS) message-passing module a message onto a queue which can be readby another process . Each message is given an identification or type so that processes can select the appropriate message. Process must share a common key in order to gain access to the queue in the first place (subject to other permissions -- see below).

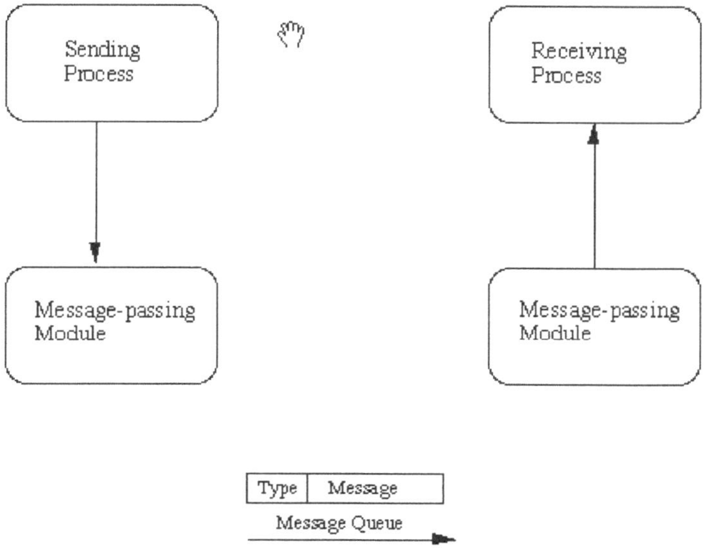

Basic Message Passing IPC messaging lets processes send and receive messages, and queue messagesfor processing in an arbitrary order. Unlike the file byte-stream data flow of pipes, each IPC message has an explicit length. Messages can be assigned a specific type. Because of this, a server process can direct message traffic between clients on its queue by using the client process PID as the message type. For single-message transactions, multiple server processes can work in parallel on transactions sent to a shared message queue.

Before a process can send or receive a message, the queue must be initialized (through the msgget function see below) Operations to send and receive messages are performed by the msgsnd() and msgrcv() functions, respectively.

When a message is sent, its text is copied to the message queue. The msgsnd() and msgrcv() functions canbe performed as either blocking or non-blocking operations. Non-blocking operations allow for asynchronous message transfer -- the process is not suspended as a result of sending or receiving a message. In blocking or synchronous message passing the sending process cannot continue until the message has been transferred or has even been acknowledged by a receiver. IPC signal and other mechanisms can be employed to implement such transfer. A blocked message operation remains suspended until one of the following three conditions occurs:

- The call succeeds.
- The process receives a signal.
- The queue is removed.

Initialising the Message Queue

The msgget() function initializes a new message queue:

int msgget(key_t key, int msgflg)

It can also return the message queue ID (msqid) of the queue corresponding to the key argument. The value passed as the msgflg argument must be an octal integer with settings for the queue's permissions and control flags.

The following code illustrates the msgget() function.

```
#include <sys/ipc.h>;

#include <sys/msg.h>;

...

key_t key; /* key to be passed to msgget()

*/ int msgflg /* msgflg to be passed to msgget()

*/int msqid; /* return value from msgget() */

...

key = ...

msgflg = ...
```

```
if ((msqid = msgget(key, msgflg)) == &ndash;1)
{
perror("msgget: msgget failed"); exit(1);
} else
(void) fprintf(stderr, &ldquo;msgget succeeded");
...
```

IPC Functions, Key Arguments, and Creation Flags: <sys/ipc.h>

Processes requesting access to an IPC facility must be able to identify it. To do this, functions that initialize or provide access to an IPC facility use a key_t key argument. (key_t is essentially an int type defined in <sys/types.h>

The key is an arbitrary value or one that can be derived from a common seed at run time. One way is withftok() , which converts a filename to a key value that is unique within the system. Functions that initialize or get access to messages (also semaphores or shared memory see later) return an ID number of type int. IPC functions that perform read, write, and control operations use this ID. If the key argument is specifiedas IPC_PRIVATE, the call initializes a new instance of an IPC facility that is private to the creating process. When the IPC_CREAT flag is supplied in the flags argument appropriate to the call, the function

tries to create the facility if it does not exist already. When called with both the IPC_CREAT and IPC_EXCL flags, the function fails if the facility already exists. This can be useful when more than one process might attempt to initialize the facility. One such case might involve several server processes having access to the same facility. If they all attempt to create the facility with IPC_EXCL in effect, only the first attempt succeeds. If neither of these flags is given and the facility already exists, the functions to get access simply return the ID of the facility. If IPC_CREAT is omitted and the facility is not already initialized, the calls fail. These control flags are combined, using logical (bitwise) OR, with the octal permission modes to form the flags argument. For example, the statement below initializes a new message queue if the queue does not exist.

msqid = msgget(ftok("/tmp",

key), (IPC_CREAT | IPC_EXCL | 0400));

The first argument evaluates to a key based on the string ("/tmp"). The second argument evaluates to the combined permissions and control flags.

Controlling message queues

The msgctl() function alters the permissions and other characteristics of a message queue. The owner or creator of a queue can change its ownership or permissions using msgctl() Also, any process with permission to do so can use msgctl() for control operations.

The msgctl() function is prototypes as follows:

int msgctl(int msqid, int cmd, struct msqid_ds *buf)

The msqid argument must be the ID of an existing message queue. The cmd argument is one of:

IPC_STAT

-- Place information about the status of the queue in the data structure pointed to by buf. The process must have read permission for this call to succeed.

IPC_SET

-- Set the owner's user and group ID, the permissions, and the size (in number of bytes) of the message queue. A process must have the effective user ID of the owner, creator, or superuser for this call to succeed.

IPC_RMID

-- Remove the message queue specified by the msqid argument.

The following code illustrates the msgctl() function with all its various flags:#include<sys/types.h>

#include <sys/ipc.h>

#include <sys/msg.h>

```
...
if (msgctl(msqid, IPC_STAT, &buf) == -1) {
perror("msgctl: msgctl failed");
exit(1);
}
...
if (msgctl(msqid, IPC_SET, &buf) == -1) {
perror("msgctl: msgctl failed");
exit(1);
}
...
```

Sending and Receiving Messages

The msgsnd() and msgrcv() functions send and receive messages, respectively:

int msgsnd(int msqid, const void *msgp, size_t msgsz,int msgflg);

int msgrcv(int msqid, void *msgp, size_t msgsz, long msgtyp, int msgflg);

The msqid argument **must** be the ID of an existing message queue. The msgp argument is a pointer to a structure that contains the type of the message and its text. The structure below is an example of what thisuser-defined buffer might look like:

```
struct mymsg {
    long        mtype;        /* message type */
    char mtext[MSGSZ]; /* message text of length MSGSZ */
}
```

The msgsz argument specifies the length of the message in bytes.

The structure member msgtype is the received message's type as specified by the sending process.The argument msgflg specifies the action to be taken if one or more of the following are true:

- The number of bytes already on the queue is equal to msg_qbytes.
- The total number of messages on all queues system-wide is equal to the system-imposed limit.

These actions are as follows:

- If (msgflg & IPC_NOWAIT) is non-zero, the message will not be sent and the calling processwill return immediately.
- If (msgflg & IPC_NOWAIT) is 0, the calling process will suspend execution until one of thefollowing occurs:
  - o The condition responsible for the suspension no longer exists, in which case the messageis sent.
  - o The message queue identifier msqid is removed from the system; when this occurs, errnois set equal to EIDRM and -1 is returned.
  - o The calling process receives a signal that is to be caught; in this case the message is notsent and the calling process resumes execution.

Upon successful completion, the following actions are taken with respect to the data structureassociated with msqid:

- o msg_qnum is incremented by 1.
- o msg_lspid is set equal to the process ID of the calling process.
- o msg_stime is set equal to the current time.

The following code illustrates msgsnd() and msgrcv():

```
#include <sys/types.h>

#include <sys/ipc.h>

#include <sys/msg.h>

...

int msgflg; /* message flags for the operation */

struct msgbuf *msgp; /* pointer to the message buffer */ int
msgsz; /* message size */
```

```
long msgtyp; /* desired message type */

int msqid /* message queue ID to be used */

...

msgp = (struct msgbuf *)malloc((unsigned)(sizeof(struct msgbuf)

- sizeof msgp->mtext + maxmsgsz));

if (msgp == NULL) {

(void) fprintf(stderr, "msgop: %s %d byte messages.\n","could
not allocate message buffer for", maxmsgsz); exit(1);

...

msgsz = ... msgflg = ...

if (msgsnd(msqid, msgp, msgsz, msgflg) == -1)perror("msgop:
msgsnd failed");

...

msgsz = ...

msgtyp = first_on_queue;msgflg = ...

if (rtrn = msgrcv(msqid, msgp, msgsz, msgtyp, msgflg) == -1)
perror("msgop: msgrcv failed");

...
```

message_send.c-- creating and sending to a simple message queue

The full code listing for message_send.cis as follows: #include <sys/types.h>

```
#include <sys/ipc.h> #include <sys/msg.h>#include <stdio.h>
#include <string.h>

#define MSGSZ          128

/*

* Declare the message structure.

*/

typedef struct msgbuf {long mtype;

char     mtext[MSGSZ];
```

```
} message_buf;
main()
{
int msqid;
int msgflg = IPC_CREAT | 0666;key_t key;
message_buf sbuf;size_t buf_length;
/*
*    Get the message queue id for the
*    "name" 1234, which was created by
*    the server.
*/
key = 1234;
(void) fprintf(stderr, "\nmsgget: Calling msgget(%#lx,\
%#o)\n", key, msgflg);
if ((msqid = msgget(key, msgflg )) < 0) {perror("msgget");
exit(1);
}
else
(void) fprintf(stderr,"msgget: msgget succeeded: msqid = %d\n",
msqid);
/*
*    We'll send message type 1
*/
sbuf.mtype = 1;
(void) fprintf(stderr,"msgget: msgget succeeded: msqid = %d\n",
msqid);(void) strcpy(sbuf.mtext, "Did you get this?");
(void) fprintf(stderr,"msgget: msgget succeeded: msqid = %d\n",
msqid);buf_length = strlen(sbuf.mtext) + 1 ;
```

```
/*
*    Send a message.
*/
if (msgsnd(msqid, &sbuf, buf_length, IPC_NOWAIT) < 0) {
printf ("%d, %d, %s, %d\n", msqid, sbuf.mtype, sbuf.mtext,
buf_length);perror("msgsnd");
exit(1);
}
else
printf("Message: \"%s\" Sent\n", sbuf.mtext);
exit(0);
}
```

The essential points to note here are:

- The Message queue is created with a basic key and message flag msgflg = IPC_CREAT | 0666

-- create queue and make it read and appendable by all.

- A message of type (sbuf.mtype) 1 is sent to the queue with the message ``Did you get this?''

message_rec.c-- receiving the above message

The full code listing for message_send.c's companion process, message_rec.cis as follows:#include <sys/types.h>

```
#include <sys/ipc.h> #include <sys/msg.h>#include <stdio.h>
#define MSGSZ                 128
/*
* Declare the message structure.
*/
typedef struct msgbuf {long mtype;
char     mtext[MSGSZ];
```

```
} message_buf;
main()
{
int msqid;key_t key;
message_buf  rbuf;
/*
*    Get the message queue id for the
*    "name" 1234, which was created by
*    the server.
*/
key = 1234;
if ((msqid = msgget(key, 0666)) < 0) {perror("msgget");
exit(1);
}
/*
*    Receive an answer of message type 1.
*/
if (msgrcv(msqid, &rbuf, MSGSZ, 1, 0) < 0) {perror("msgrcv");
exit(1);
}
/*
*    Print the answer.
*/
printf("%s\n", rbuf.mtext);exit(0);
}
```

The essential points to note here are:

- The Message queue is opened with msgget (message flag 0666) and the *same* key as message_send.c.
- A message of the *same* type 1 is received from the queue with the message ``Did you get this?''stored in rbuf.mtext.

UNIT : V

Inter Process Communication

Objective: How the two or more process communication done Semaphore

CONTENT OVERVIEW

Semaphores are not used to exchange a large amount of data. Semaphores are used synchronization among processes. Other synchronization mechanisms include record locking and mutexes. Why necessary? Examples include: shared washroom, common rail segment, and common bankaccount.

Further comments:

1) The semaphore is stored in the kernel: Allows atomic operations on the semaphore. Processes are prevented from indirectly modifying the value.

2) A process acquires the semaphore if it has a value of zero. The value of the semaphore is thenincremented to

1). When a process releases the semaphore, the value of the semaphore is decremented.

3) If the semaphore has non-zero value when a process tries to acquire it, that process blocks.

4) In comments 2 and 3, the semaphore acts as a customer counter. In most cases, it is a resource counter.

5) When a process waits for a semaphore, the kernel puts the process "to sleep" until the semaphore is available. This is better (more efficient) than busy waiting such as TEST&SET.

6) The kernel maintains information on each semaphore internally, using a data structure struct semis_ds that keeps track of permission, number of semaphores, etc.

7) Apparently, a semaphore in Unix is not a single binary value, but a set of nonnegative integer values semid ---> returned by semget

Kernel data structures for a semaphore set of 2 members

sem_perms

structure

sem_base

sem_nsems

sem_otime

struct semid_ds

sem_ctime

semval [0]:semaphore value, nonnegative

sempid [0]:pid of last successfuloperation

semzcnt [0]:# of processes awaiting semval=0

semncnt [0]:# of processes awaiting semval

semval [1]

sempid [1]

semzcnt [1]

semncnt [1]

# of semaphores

time of last semop

time of last changeres

8) There are 3 (logical) types of semaphores:

Binary semaphore - have a value of 0 or 1. Similar to a mutex lock. 0 means locked; 1 means unlocked.semaphore - has a value ≥ 0. Used for counting resources, like the producer-consumer example. Note that value =0 is similar to a lock (resource not available). Set of counting semaphores -one or more semaphores, each of which is a counting semaphore.

9) There are 2 basic operations performed with semaphores:

Wait - waits until the semaphore is > 0, then decrements it.

Post - increments the semaphore, which wakes waiting processes. Suppose you know:

concurrent process, critical region, shared resource, deadlock, mutual exclusion, primitive, atomicoperation.

semaphore set is created using :

int semget(key_t key, int nsems, int semflag); -- returns int semid; the id of the semaphore set; -1 on error.

nsems — — — — # of semaphores, use multiple semaphores for multiple resources. semflag — — — — Same as msgflag, sets permission and creation options. See

Operations on a semaphore are performed using:

int semop(int semid, struct sembuf *opsptr, unsigned int nops) semid —value returned by semget.

nops — # of operations to perform, or the number of elements in the opsptr array. opsptr — points to an array of one or more operations. Each operation is defined as:

struct sembuf { ushort sem_num; /* semaphore #, numbered from 0, 1, 2 ... */short sem_op; /* semaphore operation */

short sem_flg; /*operations flags, such as 0, IPC_NOWAIT for nonblocking call, or SEM_UNDO to have the semaphore automatically released when the

process is terminated prematurely.*/

};

sem_op = 0 - wait until the semaphore is 0. IPC_NOWAIT causes an error if semval≠0. sem_op > 0 - increment the semaphore value: semval + sem_op, (acquire)

sem_op < 0 - wait until the semaphore value≥|sem_op| and **Decrement the semaphore value**: semval - |sem_op|, (release) More notes:

As a customer counter, a semaphore is acquired doing the first two operations in one call; a semaphore isreleased using the third operation. See the following program example.

Blocking calls end when the request is satisfied, the semaphore set is deleted, or a signal is received.Keep in mind: all operations in one semop( ) must be finished atomically by the kernel. Either all or none of operations will be done.

Control operations are performed using:

int semctl(int semid, int semnum, int cmd, union semum arg); – Return value depends on cmd, -1 onerror.

union semun { int val; /* used for SETVAL only */

stuct semid_ds *buff /* used for IPC_STAT and IPC_SET */ ushort *array /* used for GETALL and SETALL*/

} arg;

Which field is used in the union depends on the cmd.

cmd --- IPC_RMID to remove a semaphore set. Union semun arg is not used in this case.

GETVAL /SETVAL to fetch/set a specific value. semnum can specify a member of the semaphore

set.

GETALL/SETALL to fetch/set all values of the semaphore set.

A semaphore set with one semaphore would be initialized using:

union semun arg;arg.val =1;

semctl(semid, 0, SETVAL, arg);

Example: How to write lock/unlock (somewhat like P/V operations)#include <sys/types.h>

```
#include <sys/ipc.h> #include <sys/sem.h>

#define SEMKEY 123456L /* key value for semget() */
#define PERMS 0666

static struct sembuf op_lock[2]= { 0, 0, 0,
/* wait for sem #0 to become 0 */ 0, 1, SEM_UNDO /* then
increment sem #0 by 1 */ };

static struct sembuf op_unlock[1]= { 0, -1, (IPC_NOWAIT |
SEM_UNDO)

/* decrement sem #0 by 1 (sets it to 0) */ };

int semid = -1; /* semaphore id. Only the first time will create
a semaphore.*/

my_lock( )

{ if (semid <0) {

if ( ( semid=semget(SEMKEY, 1, IPC_CREAT | PERMS )) < 0 )
printf("semget error"); }if (semop(semid, &op_lock[0], 2) < 0)
printf("semop lock error");

}

my_unlock( )

{

if (semop(semid, &op_unlock[0], 1) < 0) printf("semop unlock
error");

}
```

Questions: how to rewrite the above program to make the semaphore as a resource counter? What if theresource allows 3 or more processes to use at the same time?

Solutions:

```
static struct sembuf op_lock[1]= { 0, -1, SEM_UNDO, };

static struct sembuf op_unlock[1]= { 0, 1, (IPC_NOWAIT |
SEM_UNDO) }
```

Don't forget to set the initial value of the semaphore as 1 or 3.

Shared Memory

Using a pipe or a message queue requires multiple exchanges of data through the kernel.

Shared memory can be used to bypass the kernel for faster processing.

The kernel maintains information about each shared memory segment, including permission, size, accesstime, etc

in struct shmid_ds .

struct shmid_ds looks like struct semid_ds or as struct msqid_ds .

A share memory segment is created using:

int shmget(key_t key, int size, int shmflag);

size --- size of the shared memory segment in bytes.

shmflag --- same as for msgget() and semget(), see Lecture 4's Comments 1.Return value --- shmid, the shared memory identifier, -1 on error.

Attach to the shared memory segment using:

char *shmat(int shmid, char *shmaddr, int shmflag)

shmid --- return value of shmget, that is, the id of the created shared memory.shmaddr--- 0: let the kernel select the address.

shmflag--- SHM_RDONLY for read_only access.

returns the starting address of the shared memory, and thus we can read/write on the shared memory after getting its starting address.

Detach the shared memory segment using:

char *shmdt(char *shmaddr)

shmaddr --- the return value of shmat(), that is, the starting address of the shared memory.returns -1 on failure.

To remove a shared memory segment:

Int shmctl(int shmid, int cmd, shmid_ds *buf);

cmd--- IPC_RMID to delete, e.g., shmctl(shmid, IPC_RMID, 0) .

Difference between Process and Threads Process

- An executing instance of a program is called a process.
- Some operating systems use the term 'task' to refer to a program that is being executed.
- A process is always stored in the main memory also termed as the primary memory or randomaccess memory.
- Therefore, a process is termed as an active entity. It disappears if the machine is rebooted.
- Several process may be associated with a same program.
- On a multiprocessor system, multiple processes can be executed in parallel.
- On a uni-processor system, though true parallelism is not achieved, a process scheduling algorithm is applied and the processor is scheduled to execute each process one at a time yielding anillusion of concurrency.
- **Example:** Executing multiple instances of the 'Calculator' program. Each of the instances aretermed as a process.

Thread

- A thread is a subset of the process.
- It is termed as a 'lightweight process', since it is similar to a real process but executes within the context of a process and shares the same resources allotted to the process by the kernel (See kquest.co.cc/2010/03/operating-system for more info on the term 'kernel').
- Usually, a process has only one thread of control – one set of machine instructions executing at a time.
- A process may also be made up of multiple threads of execution that execute instructions concurrently.
- Multiple threads of control can exploit the true parallelism possible on multiprocessor systems.
- On a uni-processor system, a thread scheduling algorithm is applied and the processor is scheduled to run each thread one at a time.

- All the threads running within a process share the same address space, file descriptor, stack and other process related attributes.
- Since the threads of a process share the same memory, synchronizing the access to the shared data withing the process gains unprecedented importance.

Linux Threads and Uses

To invoke threads a data structure is utilized.

Thread Data Types

- pthread_t
- pthread_mutex_t - Mutex
- pthread_cond_t - Condition variable
- pthread_key_t - Access key for thread data.
- pthread_attr_t - Thread attributes
- pthread_mutexattr_t - Mutex attributes
- pthread_condattr_t - Condition variable attributes
- pthread_once_t - One time initialization

Thread Structure and Thread Functions

1. pthread_t
2. int pthread_create (pthread_t *thread, const pthread_attr_t *attr, void *(*start)(void *), void *arg); - The thread identifier which is needed to do anything with the thread is the value returnedin the first argument, *thread . The third argument is the address of the thread routine to run.
3. pthread_t pthread_self (void);
4. int pthread_detach (pthread_t thread); - Allows the system resources for the thread to be releasedwhen the thread exits.
5. int pthread_join (pthread_t thread, void **value_ptr); - Blocks until the thread specifiedterminates. It will optionally store the return value of the terminated thread.

6. int pthread_exit (void *value_ptr);
7. int pthread_equal (pthread_t thr1, pthread_t thr2); - Returns 0 value if the threads are not equaland non-zero if they are the same thread.
8. pthread_t pthread_self (void); - Allows a thread to get its own identifier.

Thread States

1. ready - Ready to run in the system scheduler.
2. blocked - Waiting for a mutex or resource.
3. running - Running by the system scheduler
4. terminated - The thread has normally exited or has called Pthread_exit to exit. Its resources havenot been freed and will be freed if it is detached or joined.

Thread Identification

Just as a process is identified through a process ID, a thread is identified by a thread ID. Butinterestingly, the similarity between the two ends here.

- A process ID is unique across the system where as a thread ID is unique only in context of a single process.
- A process ID is an integer value but the thread ID is not necessarily an integer value. It could well be a structure
- A process ID can be printed very easily while a thread ID is not easy to print.

The above points give an idea about the difference between a process ID and thread ID.

Thread ID is represented by the type 'pthread_t'. As we already discussed that in most of thecases this type is a structure, so there has to be a function that can compare two thread IDs.

```
#include <pthread.h>

int pthread_equal(pthread_t tid1, pthread_t tid2);
```

So as you can see that the above function takes two thread IDs and returns nonzero value if boththe thread IDs are equal or else it returns zero.

Another case may arise when a thread would want to know its own thread ID. For this case thefollowing function provides the desired service.

#include <pthread.h> pthread_t pthread_self(void);

So we see that the function 'pthread_self()' is used by a thread for printing its own thread ID.

Now, one would ask about the case where the above two function would be required. Suppose there is a case where a link list contains data for different threads. Every node in the list containsa thread ID and the corresponding data. Now whenever a thread tries to fetch its data from linkedlist, it first gets its own ID by calling 'pthread_self()' and then it calls the 'pthread_equal()' on every node to see if the node contains data for it or not.

An example of the generic case discussed above would be the one in which a master thread getsthe jobs to be processed and then it pushes them into a link list. Now individual worker threads parse the linked list and extract the job assigned to them.

Thread Creation

Normally when a program starts up and becomes a process, it starts with a default thread. So wecan say that every process has at least one thread of control. A process can create extra threads using the following function :

#include <pthread.h>

int pthread_create(pthread_t *restrict tidp, const pthread_attr_t *restrict attr, void

*(*start_rtn)(void), void *restrict arg)

The above function requires four arguments, lets first discuss a bit on them :

- The first argument is a pthread_t type address. Once the function is called successfully, the variable whose address is passed as first argument will hold the thread ID of the newly created thread.

- The second argument may contain certain attributes which we want the new thread to contain. It could be priority etc.
- The third argument is a function pointer. This is something to keep in mind that each thread starts with a function and that functions address is passed here as the third argument so that the kernel knows which function to start the thread from.
- As the function (whose address is passed in the third argument above) may accept some arguments also so we can pass these arguments in form of a pointer to a void type. Now, why a void type was chosen? This was because if a function accepts more than one argument then this pointer could be a pointer to a structure that may contain these arguments.

Thread Example

Following is the example code where we tried to use all the three functions discussed above.

```
#include<stdio.h>

#include<string.h>

#include<pthread.h>

#include<stdlib.h>

#include<unistd.h>

pthread_t tid[2];

void* doSomeThing(void *arg)

{

    unsigned long i = 0;

    pthread_t id = pthread_self();

    if(pthread_equal(id,tid[0]))

    {

    printf("\n First thread processing\n");
```

```
}
else
{
printf("\n Second thread processing\n");
}
for(i=0; i<(0xFFFFFFFF);i++);
return NULL;
}
int main(void)
{
int i = 0;
int err; while(i < 2)
{
err = pthread_create(&(tid[i]), NULL, &doSomeThing, NULL);
if (err != 0)
printf("\ncan't create thread :[%s]", strerror(err));else
printf("\n Thread created successfully\n");
i++;
}
sleep(5);return 0;
}
```

So what this code does is :

- It uses the pthread_create() function to create two threads
- The starting function for both the threads is kept same.
- Inside the function 'doSomeThing()', the thread uses pthread_self() and pthread_equal() functions to identify whether the executing thread is the first one or the second one as created.
- Also, Inside the same function 'doSomeThing()' a for loop is run so as to simulate some time consuming work.

output :

$ ./threads

Thread created successfullyFirst thread processing Thread created successfullySecond thread processing

Heavyweight vs Lightweight Processes

As we have seen above, Xinu processes all execute in the same address space, and do not incur the overhead of switching to the kernel address space during context switching. Such processes are called **lightweight processes** (**lwps**), since they have little associated state and share memory with each other and the process manager, making context switches, process creation, and interprocess communication relatively inexpensive. These processes are to contrasted with Unix-like **heavyweight processes** (**hwps**), which run in separate address spaces and switch to the kernel address space on context switches. Lightweight and heavyweight processes are complementary concepts in that one can run multiple lightweight processes inside a heavyweight

process. In fact, your assignments will be doing exactly this, creating Xinu lwps within a Unixhwp.

Thread Attributes:

- By default, a thread is created with certain attributes. Some of these attributes can be changed by the programmer via the thread attribute object.
- pthread_attr_init and pthread_attr_destroy are used to initialize/destroy the thread attribute object.
- Other routines are then used to query/set specific attributes in the thread attribute object. Attributes include:
    - Detached or joinable state
    - Scheduling inheritance
    - Scheduling policy
    - Scheduling parameters
    - Scheduling contention scope

- Stack size
- Stack address
- Stack guard (overflow) size

POSIX Thread:

- Process ID, process group ID, user ID, and group ID
- Environment
- Working directory.
- Program instructions
- Registers
- Stack
- Heap
- File descriptors
- Signal actions
- Shared libraries
- Inter-process communication tools (such as message queues, pipes, semaphores, orshared memory).

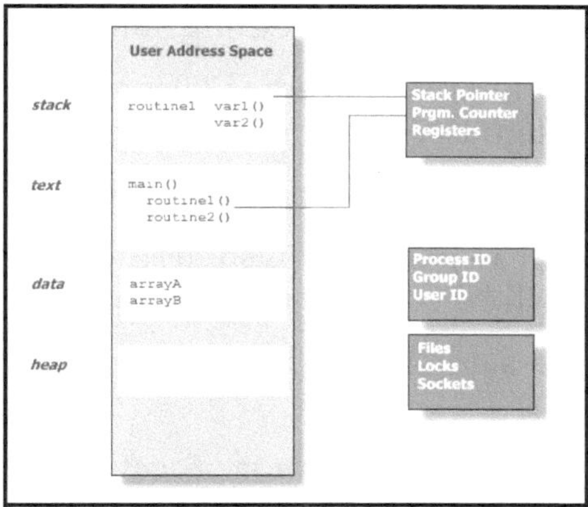

LINUX PROCESS

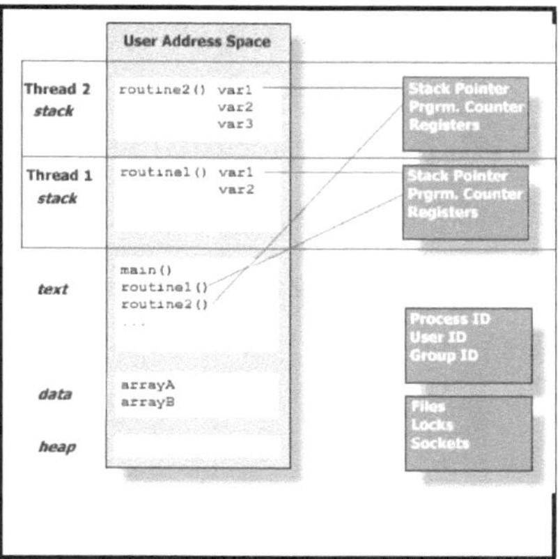

THREADS WITHIN A LINUX PROCESS

- Threads use and exist within these process resources, yet are able to be scheduled by the operating system and run as independent entities largely because they duplicate only the bare essential resources that enable them to exist as executable code.

- This independent flow of control is accomplished because a thread maintains its own:
  - Stack pointer
  - Registers
  - Scheduling properties (such as policy or priority)
  - Set of pending and blocked signals
  - Thread specific data.
- So, in summary, in the UNIX environment a thread:
  - Exists within a process and uses the process resources
  - Has its own independent flow of control as long as its parent process exists and the OS supports it
  - Duplicates only the essential resources it needs to be independently schedulable
  - May share the process resources with other threads that act equally independently (and dependently)
  - Dies if the parent process dies - or something similar
  - Is "lightweight" because most of the overhead has already been accomplished through the creation of its process.
- Because threads within the same process share resources:
  - Changes made by one thread to shared system resources (such as closing a file) will be seen by all other threads.
  - Two pointers having the same value point to the same data.
  - Reading and writing to the same memory locations is possible, and therefore requires

The Pthreads API

- The original Pthreads API was defined in the ANSI/IEEE POSIX 1003.1 - 1995 standard. The POSIX standard has continued to evolve and undergo revisions, including the Pthreads specification.

- Copies of the standard can be purchased from IEEE or downloaded for free from other sites online.
- The subroutines which comprise the Pthreads API can be informally grouped into four major groups:
  1. **Thread management:** Routines that work directly on threads - creating, detaching, joining, etc. They also include functions to set/query thread attributes (joinable, scheduling etc.)
  2. **Mutexes:** Routines that deal with synchronization, called a "mutex", which is an abbreviation for "mutual exclusion". Mutex functions provide for creating, destroying, locking and unlocking mutexes. These are supplemented by mutex attribute functions thatset or modify attributes associated with mutexes.
  3. **Condition variables:** Routines that address communications between threads that share a mutex. Based upon programmer specified conditions. This group includes functions to create, destroy, wait and signal based upon specified variable values. Functions to set/query condition variable attributes are also included.
  4. **Synchronization:** Routines that manage read/write locks and barriers.
- Naming conventions: All identifiers in the threads library begin with pthread_. Some examples are shown below.

Routine Prefix	Functional Group
pthread_	Threads themselves and miscellaneous subroutines
pthread_attr_	Thread attributes objects
pthread_mutex_	Mutexes
pthread_mutexattr_	Mutex attributes objects.
pthread_cond_	Condition variables
pthread_condattr_	Condition attributes objects

pthread_key_	Thread-specific data keys
pthread_rwlock_	Read/write locks
pthread_barrier_	Synchronization barriers

- The concept of opaque objects pervades the design of the API. The basic calls work to create or modify opaque objects - the opaque objects can be modified by calls to attribute functions, which deal with opaque attributes.
- The Pthreads API contains around 100 subroutines. This tutorial will focus on a subset of these - specifically, those which are most likely to be immediately useful to the beginning Pthreads programmer.
- For portability, the pthread.h header file should be included in each source file using the Pthreads library.
- The current POSIX standard is defined only for the C language. Fortran programmers can use wrappers around C function calls. Some Fortran compilers (like IBM AIX Fortran) may provide a Fortram pthreads API.
- A number of excellent books about Pthreads are available. Several of these are listed in the References section of this tutorial.

Thread Management

Creating and Terminating Threads

Routines:

pthread_create (thread,attr,start_routine,arg)

pthread_exit (status)

pthread_cancel (thread)

pthread_attr_init (attr)

pthread_attr_destroy (attr)

Creating Threads:

- Initially, your main() program comprises a single, default thread. All other threads must be explicitly created by the programmer.
- pthread_create creates a new thread and makes it executable. This routine can be called any number of times from anywhere within your code.
- pthread_create arguments:
  - o thread: An opaque, unique identifier for the new thread returned by the subroutine.
  - o attr: An opaque attribute object that may be used to set thread attributes. You can specify a thread attributes object, or NULL for the default values.
  - o start_routine: the C routine that the thread will execute once it is created.
  - o arg: A single argument that may be passed to *start_routine*. It must be passed by reference as a pointer cast of type void. NULL may be used if no argument is to be passed.
- The maximum number of threads that may be created by a process is implementation dependent.
- Once created, threads are peers, and may create other threads. There is no implied hierarchy or dependency between threads.

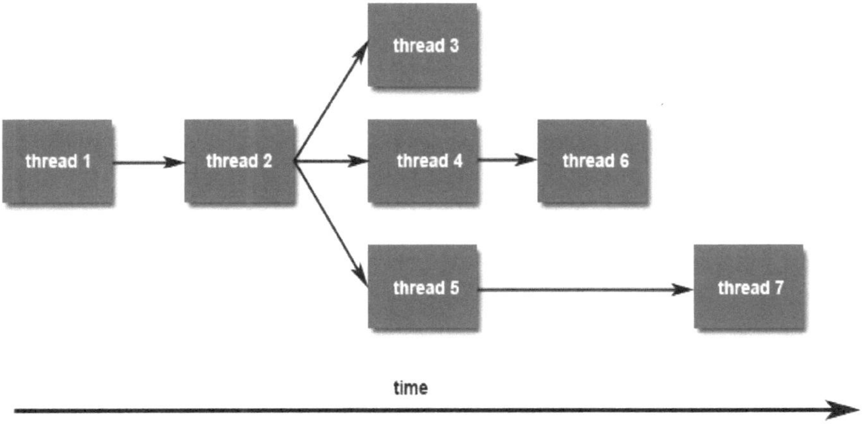

Thread Attributes:

- By default, a thread is created with certain attributes. Some of these attributes can be changed by the programmer via the thread attribute object.
- pthread_attr_init and pthread_attr_destroy are used to initialize/destroy the thread attribute object.
- Other routines are then used to query/set specific attributes in the thread attribute object. Attributes include:
  - Detached or joinable state
  - Scheduling inheritance
  - Scheduling policy
  - Scheduling parameters
  - Scheduling contention scope
  - Stack size
  - Stack address
  - Stack guard (overflow) size
- Some of these attributes will be discussed later.

Thread Binding and Scheduling:

- The Pthreads API provides several routines that may be used to specify how threads are scheduled for execution. For example, threads can be scheduled to run FIFO (first-in first-out), RR (round-robin) or OTHER (operating system determines). It also provides the ability to set a thread's scheduling priority value.
- These topics are not covered here, however a good overview of "how things work" under Linux can be found in the sched_setscheduler man page.
- The Pthreads API does not provide routines for binding threads to specific cpus/cores. However, local implementations may include this functionality - such as providing the non-standard pthread_setaffinity_np routine. Note that "_np" in the name stands for "non-portable".

- Also, the local operating system may provide a way to do this. For example, Linux provides the sched_setaffinity routine.

Terminating Threads & pthread_exit():

- There are several ways in which a thread may be terminated:
    - The thread returns normally from its starting routine. It's work is done.
    - The thread makes a call to the pthread_exit subroutine - whether its work is done or not.
    - The thread is canceled by another thread via the pthread_ cancel routine.
    - The entire process is terminated due to making a call to either the exec() or exit()
    - If main() finishes first, without calling pthread_exit explicitly itself
- The pthread_exit() routine allows the programmer to specify an optional termination status parameter. This optional parameter is typically returned to threads "joining" the terminated thread (covered later).
- In subroutines that execute to completion normally, you can often dispense with calling pthread_exit() - unless, of course, you want to pass the optional status code back.
- Cleanup: the pthread_exit() routine does not close files; any files opened inside the thread will remain open after the thread is terminated.
- Discussion on calling pthread_exit() from main():
    - There is a definite problem if main() finishes before the threads it spawned if you don't call pthread_exit() explicitly. All of the threads it created will terminate because main() is done and no longer exists to support the threads.
    - By having main() explicitly call pthread_exit() as the last thing it does, main() willblock and be kept alive to support the threads it created until they are done.

Example: Pthread Creation and Termination

```
#include <pthread.h>
#include <stdio.h>
#define NUM_THREADS     5

void *PrintHello(void *threadid)
{
   long tid;
   tid – (long)threadid;
   printf("Hello World! It's me, thread #%ld!\n", tid);
   pthread_exit(NULL);
}

int main (int argc, char *argv[])
{
   pthread_t threads[NUM_THREADS];
   int rc;
   long t;
   for(t=0; t<NUM_THREADS; t++){
     printf("In main: creating thread %ld\n", t);
     rc – pthread_create(&threads[t], NULL, PrintHello, (void *)t);
     if (rc){
       printf("ERROR; return code from pthread_create() is %d\n", rc);
       exit(-1);
     }
   }

   /* Last thing that main() should do */
   pthread_exit(NULL);
}
```

Thread Synchronization with Semaphores with Mutexes

Introduction

Semaphore is a synchronization technique where we can control number of threads to access aresource. In lock/Mutex, only one thread can access resources at a time. But Semaphore allows multiple threads to access the same resource at a time. We can limit the number of threads thatcan access the same resources. In this article, I have shown three different ways to access resources.

1. No Synchronization
2. Synchronization with Monitor

3. Synchronization with Semaphore

1. No Synchronization

With no synchronization, all threads run simultaneously and execute the same piece of code simultaneously. There is no restriction on how many threads can access it. Following is the code:

```
private void btnNoSync_Click(object sender, EventArgs e)

{

listBox1.Items.Add("== No Synchronization ==========");
int TotalThread = 5;

Thread[] Threads = new Thread[TotalThread];for (int i = 0; i <
TotalThread; i++)

{

Threads[i] = new Thread(new ThreadStart(AccessCode));Threads[i].
IsBackground = true;

Threads[i].Start();

}

}

public void AccessCode()

{

listBox1.BeginInvoke(new ParameterizedThreadStart(Update
UI), new object[]

{"Thread ID : " + Thread.CurrentThread.ManagedThreadId.
ToString() + ": Entered" }

);

Thread.Sleep(500);

listBox1.BeginInvoke(new ParameterizedThreadStart(Update
UI), new object[]

{ "Thread ID : " + Thread.CurrentThread.ManagedThreadId.
ToString() + " : Exit" }
```

```
);
}
// Following code is used to update UI public void UpdateUI(object objOutput)
{
listBox1.Items.Add(objOutput.ToString());
}
```

2. Synchronization with Monitor

Synchronization with monitor class, only one thread can access the same resource at a time.Thread is run simultaneously but it can access the block of code one at a time. There is a restriction on thread so that only a single thread can access a particular code block.

```
private void buttonMonitor_Click(object sender, EventArgs e)
{
listBox1.Items.Add("== Using Monitor =============");int TotalThread = 5;
Thread[] Threads = new Thread[TotalThread];for (int i = 0; i < TotalThread; i++)
{
Threads[i] = new Thread(new ThreadStart(AccessCodeWithMonitor));Threads[i].IsBackground = true;
Threads[i].Start();
}
}
private void AccessCodeWithMonitor()
{
Monitor.Enter(this);try
{
```

```
listBox1.BeginInvoke(new ParameterizedThreadStart(Update
UI), new object[]
{ "Thread ID : " +
Thread.CurrentThread.ManagedThreadId.ToString() + " : Entered"
});Thread.Sleep(500);
listBox1.BeginInvoke(new ParameterizedThreadStart(Update
UI), new object[]
{ "Thread ID : " +
Thread.CurrentThread.ManagedThreadId.ToString() + " : Exit" });
}
finally
{
Monitor.Exit(this);
}
}
private void UpdateUI(object objOutput)
{
listBox1.Items.Add(objOutput.ToString());
}
```

3. Synchronization with Semaphore

In synchronization with semaphore class, we can allow more than one thread to access the same block of code. Actually, we can specify how many threads can access the same block of code atthe same time.

```
private void btnSemaphore_Click(object sender, EventArgs e)
{
listBox1.Items.Add("== Using Semaphore =============");
int TotalThread = 5;
int SemaphoreCount = 3;
```

```
Thread[] Threads = new Thread[TotalThread];
Semaphore Sema = new Semaphore(SemaphoreCount, SemaphoreCount);for (int i = 0; i < TotalThread; i++)
{
Threads[i] = new Thread(new ParameterizedThreadStart
(AccessCodewithSemaphore));Threads[i].IsBackground = true;
Threads[i].Start(Sema);
}
}
public void AccessCodewithSemaphore(object objSemaphore)
{
bool IsComplete = false;
Semaphore l_SemaPhore = (Semaphore)objSemaphore; while (!IsComplete)
{
if (l_SemaPhore.WaitOne(200, false))
{
try
{
listBox1.BeginInvoke(new ParameterizedThreadStart(UpdateUI),
new object[]
{"Thread ID : " +
Thread.CurrentThread.ManagedThreadId.ToString() + " : Entered" });Thread.Sleep(500);
}
finally
{
l_SemaPhore.Release();
```

```
listBox1.BeginInvoke(new ParameterizedThreadStart(UpdateUI),
new object[]
{ "Thread ID : " +
Thread.CurrentThread.ManagedThreadId.ToString() + " : Exit"
});IsComplete = true;
}
}
else
{
listBox1.BeginInvoke(new ParameterizedThreadStart(UpdateUI),
new object[]
{"Thread ID :"+Thread.CurrentThread.ManagedThreadId.
ToString() +
": Waiting To enter"});
}
}
}
public void UpdateUI(object objOutput)
{
listBox1.Items.Add(objOutput.ToString());
}
```

Socket Programming

Objective: This Unit gives introduction about network programming and discusses about socket creationin connection oriented and connection less network.

CONTENT OVERVIEW

Socket creation introduction:

Sockets allow communication between two different processes on the same or different machines. To be more precise, it's a way to talk to other computers using standard Unix file descriptors. In Unix, every I/O actions are done by writing or reading to a file descriptor. A file descriptoris just an integer associated with an open file and it can be a network connection, a text file, a terminal, or something else.

To a programmer a socket looks and behaves much like a low level file descriptor. This is because commands such as read() and write() work with sockets in the same way they do with files and pipes. Thedifferences between sockets and normal file descriptors occurs in the creation of a socket and through a variety of special operations to control a socket.

Sockets were first introduced in 2.1BSD and subsequently refined into their current form with 4.2BSD. The sockets feature is now available with most current UNIX system releases.

Where is Socket used?

A Unix Socket is used in a client server application frameworks. A server is a process which does some function on request from a client. Most of the application level protocols like FTP, SMTP and POP3 make use of Sockets to establish connection between client and server and then for exchanging data.

Socket Types:

There are four types of sockets available to the users. The first two are most commenly used and last two are rarely used.

Processes are presumed to communicate only between sockets of the same type but there is no restriction that prevents communication between sockets of different types.

- Stream Sockets: Delivery in a networked environment is guaranteed. If you send through the stream socket three items "A,B,C", they will arrive in the same order - "A,B,C". These sockets use TCP (Transmission Control Protocol) for data transmission. If delivery is impossible, the sender receives an error indicator. Data records do no have any boundaries.
- Datagram Sockets: Delivery in a networked environment is not guaranteed. They're connectionless because you don't

need to have an open connection as in Stream Sockets - you build a packet with the destination information and send it out. They use UDP (User Datagram Protocol).

- Raw Sockets: provides users access to the underlying communication protocols which support socket abstractions. These sockets are normally datagram oriented, though their exact characteristics are dependent on the interface provided by the protocol. Raw sockets are not intended for the general user; they have been provided mainly for those interested in developing new communication protocols, or for gaining access to some of the more esoteric facilities of an existing protocol.
- Sequenced Packet Sockets: They are similar to a stream socket, with the exception that record boundaries are preserved. This interface is provided only as part of the Network Systems (NS) socket abstraction, and is very important in most serious NS applications. Sequenced-packet sockets allow the user to manipulate the Sequence Packet Protocol (SPP) or Internet Datagram Protocol (IDP) headers on a packet or a group of packets either by writing a prototype header along with whatever data is to be sent, or by specifying a default header to be used with all outgoing data, and allows the user to receive the headers on incoming packets.

Most of the Net Applications use the Client Server architecture. These terms refer to the two processes or two applications which will be communicating with each other to exchange some information. One of the two processes acts as a client process and another process acts as a server.

Client Process:

This is the process which typically makes a request for information. After getting the response this process may terminate or may do some other processing.

For example: Internet Browser works as a client application which sends a request to Web Server to get one HTML web page.

Server Process:

This is the process which takes a request from the clients. After getting a request from the client, this process will do required processing and will gather requested information and will send it to the requestorclient. Once done, it becomes ready to serve another client. Server process are always alert and ready to serve incoming requests.

For example: Web Server keeps waiting for requests from Internet Browsers and as soon as it gets any request from a browser, it picks up a requested HTML page and sends it back to that Browser.

Notice that the client needs to know of the existence and the address of the server, but the server does not need to know the address or even the existence of the client prior to the connection being established. Once a connection is established, both sides can send and receive information.

2- tier and 3-tier architectures:

There are two types of client server architectures:

- 2-tier architectures: In this architecture, client directly interact with the server. This type of architecture may have some security holes and performance problems. Internet Explorer and WebServer works on two tier architecture. Here security problems are resolved using Secure Socket Layer(SSL).
- 3-tier architectures: In this architecture, one more software sits in between client and server. This middle software is called middleware. Middleware are used to perform all the security checks andload balancing in case of heavy load. A middleware takes all requests from the client and after doing required authentication it passes that request to the server. Then server does required processing and sends response back to the middleware and finally middleware passes this response back to the client. If you want to implement a 3-tier architecture then you can keep any middle ware like Web Logic or WebSphere software in between your Web Server and Web Browsers.

Types of Server:

There are two types of servers you can have:

- Iterative Server: This is the simplest form of server where a server process serves one client and after completing first request then it takes request from another client. Meanwhile another client keeps waiting.
- Concurrent Servers: This type of server runs multiple concurrent processes to serve many request at a time. Because one process may take longer and another client can not wait for so long. The simplest way to write a concurrent server under Unix is to *fork* a child process to handle each client separately.

How to make client:

The system calls for establishing a connection are somewhat different for the client and the server, butboth involve the basic construct of a socket. The two processes each establish their own sockets.

The steps involved in establishing a socket on the client side are as follows:

1. Create a socket with the *socket()* system call.
2. Connect the socket to the address of the server using the *connect()* system call.
3. Send and receive data. There are a number of ways to do this, but the simplest is to use the *read()*

and *write()* system calls.

How to make a server:

The steps involved in establishing a socket on the server side are as follows:

1. Create a socket with the *socket()* system call.
2. Bind the socket to an address using the *bind()* system call. For a server socket on the Internet, anaddress consists of a port number on the host machine.

3. Listen for connections with the *listen()* system call.
4. Accept a connection with the *accept()* system call. This call typically blocks until a clientconnects with the server.
5. Send and receive data using the *read()* and *write()* system calls.

Client and Server Interaction:

Following is the diagram showing complete Client and Server interaction:

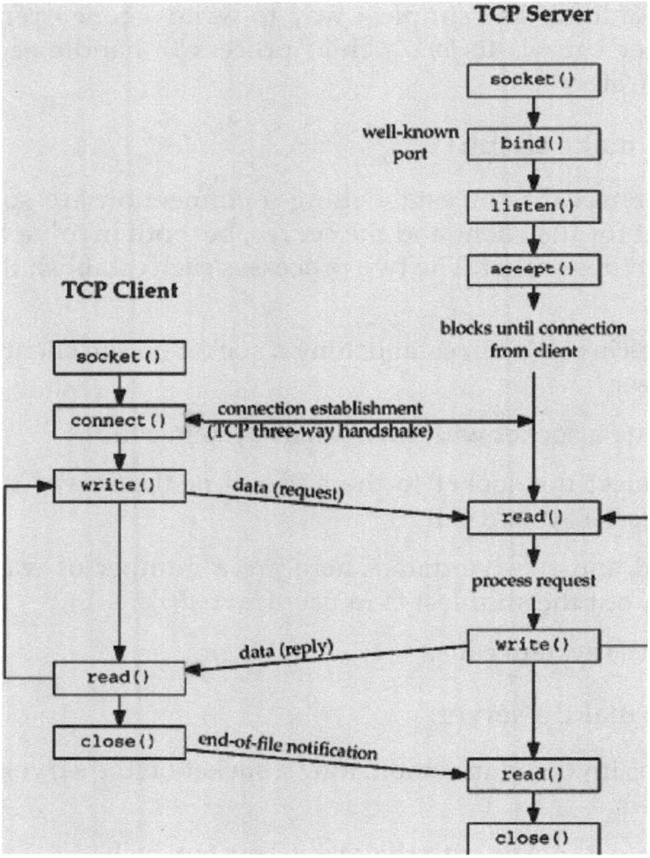

This tutorial describes the core socket functions required to write a complete TCP client and server.Following is the diagram showing complete Client and Server interaction:

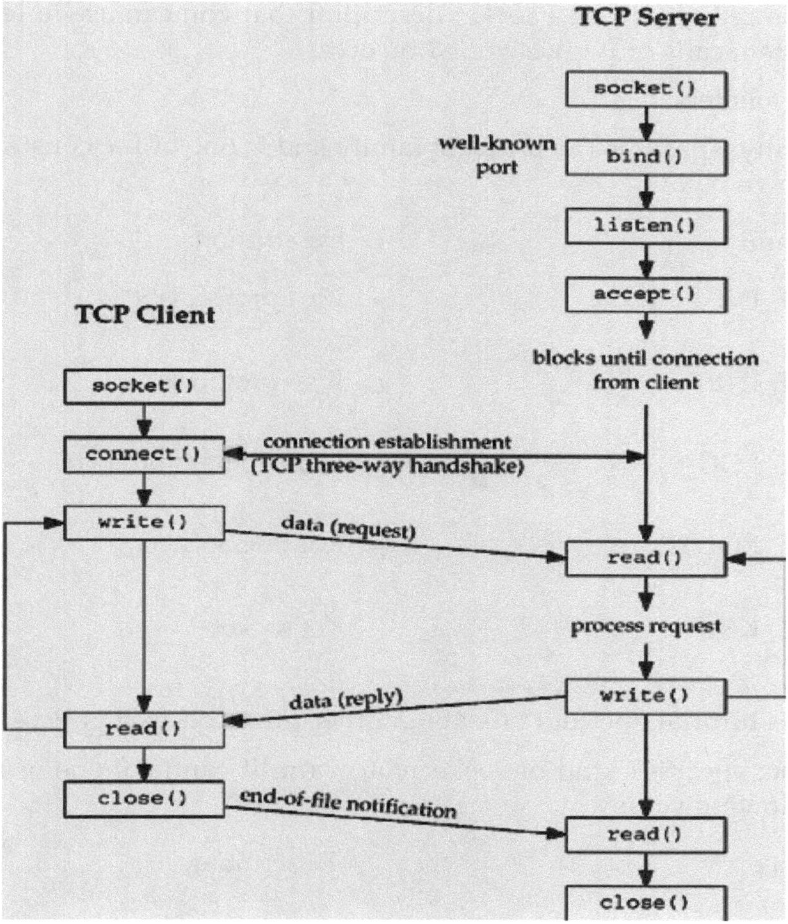

The *socket* Function:

To perform network I/O, the first thing a process must do is call the socket function, specifying the type of communication protocol desired and protocol family etc.

```
#include <sys/types.h>

#include <sys/socket.h>



int socket (int family, int type, int protocol);
```

This call gives you a socket descriptor that you can use in later system calls or it gives you -1 on error.

Parameters:

family: specifies the protocol family and is one of the constants shown below:

Family	Description
AF_INET	IPv4 protocols
AF_INET6	IPv6 protocols
AF_LOCAL	Unix domain protocols
AF_ROUTE	Routing Sockets
AF_KEY	Ket socket

This tutorial does not talk about other protocols except IPv4.

type: specifies kind of socket you want. It can take one of the following values:

Type	Description
SOCK_STREAM	Stream socket
SOCK_DGRAM	Datagram socket
SOCK_SEQPACKET	Sequenced packet socket
SOCK_RAW	Raw socket

protocol: argument should be set to the specific protocol type given below or 0 to select the system's default for the given combination of family and type:

Protocol	Description
IPPROTO_TCP	TCP transport protocol
IPPROTO_UDP	UDP transport protocol
IPPROTO_SCTP	SCTP transport protocol

The *connect* Function:

The *connect* function is used by a TCP client to establish a connection with a TCP server.

```
#include <sys/types.h>
#include <sys/socket.h>

int connect(int sockfd, struct sockaddr *serv_addr, int addrlen);
```

This call returns 0 if it successfully connects to the server otherwise it gives you -1 on error. Parameters:

- sockfd: is a socket descriptor returned by the socket function.
- serv_addr is a pointer to struct sockaddr that contains destination IP address and port.
- addrlen set it to sizeof(struct sockaddr).

The bind Function:

The *bind* function assigns a local protocol address to a socket. With the Internet protocols, the protocol address is the combination of either a 32-bit IPv4 address or a 128-bit IPv6 address, along with a 16-bit TCP or UDP port number. This function is called by TCP server only.

```
#include <sys/types.h>
#include <sys/socket.h>

int bind(int sockfd, struct sockaddr *my_addr,int addrlen);
```

This call returns 0 if it successfully binds to the address otherwise it gives you -1 on error.

Parameters:

- sockfd: is a socket descriptor returned by the socket function.
- my_addr is a pointer to struct sockaddr that contains local IP address and port.
- addrlen set it to sizeof(struct sockaddr).

You can put your IP address and your port automatically:

A 0 value for port number means system will choose a random port and *INADDR_ANY* value for IPaddress means server's IP address will be assigned automatically.

```
server.sin_port = 0;

server.sin_addr.s_addr = INADDR_ANY;
```

NOTE: As descript in Ports and Services tutorials, all ports bellow 1024 are reserved. So you can set aport above 1024 and bellow 65535 unless the ones being used by other programs.

The *listen* Function:

The *listen* function is called only by a TCP server and it performs two actions:

- The listen function converts an unconnected socket into a passive socket, indicating that the kernel should accept incoming connection requests directed to this socket.
- The second argument to this function specifies the maximum number of connections the kernel should queue for this socket.

```
#include <sys/types.h>
#include <sys/socket.h>

int listen(int sockfd,int backlog);
```

This call returns 0 on success otherwise it gives you -1 on error.

Parameters:

- sockfd: is a socket descriptor returned by the socket function.
- backlog is the number of allowed connections.

The *accept* Function:

The *accept* function is called by a TCP server to return the next completed connection from the front ofthe completed connection queue. If the completed connection queue is empty, the process is put to sleep.

- The listen function converts an unconnected socket into a passive socket, indicating that the kernel should accept incoming connection requests directed to this socket.
- The second argument to this function specifies the maximum number of connections the kernelshould queue for this socket.

```
#include <sys/types.h>
#include <sys/socket.h>

int accept (int sockfd, struct sockaddr *cliaddr, socklen_t *addrlen);
```

This call returns non negative descriptor on success otherwise it gives you -1 on error. The returned decriptor is assumed to be a client socket descriptor and all read write operations will be done on this descripton to communicate with the client.

Parameters:

- sockfd: is a socket descriptor returned by the socket function.
- cliaddr is a pointer to struct sockaddr that contains client IP address and port.

- addrlen set it to sizeof(struct sockaddr).

The send Function:

The *send* function is used to send data over stream sockets or CONNECTED datagram sockets. If you want to send data over UNCONNECTED datagram sockets you must use sendto() function.

You can use *write()* system call to send the data. This call is explained in helper functions tutorial.

```
int send(int sockfd, const void *msg, int len, int flags);
```

This call returns the number of bytes sent out otherwise it will return -1 on error.

Parameters:

- sockfd: is a socket descriptor returned by the socket function.
- msg is a pointer to the data you want to send.
- len is the length of the data you want to send (in bytes).
- flags is set to 0.

The recv Function:

The *recv* function is used to receive data over stream sockets or CONNECTED datagram sockets. If youwant to receive data over UNCONNECTED datagram sockets you must use recvfrom().

You can use *read()* system call to read the data. This call is explained in helper functions tutorial.

```
int recv(int sockfd, void *buf, int len, unsigned int flags);
```

This call returns the number of bytes read into the buffer otherwise it will return -1 on error.

Parameters:

- sockfd: is a socket descriptor returned by the socket function.
- buf is the buffer to read the information into.
- len is the maximum length of the buffer.

- flags is set to 0.

The sendto Function:

The *sendto* function is used to send data over UNCONNECTED datagram sockets. Put simply, when you use scoket type as SOCK_DGRAM

```
int sendto(int sockfd, const void *msg, int len, unsigned int flags,
                    const struct sockaddr *to, int tolen);
```

This call returns the number of bytes sent otherwise it will return -1 on error.

Parameters:

- sockfd: is a socket descriptor returned by the socket function.
- msg is a pointer to the data you want to send.
- len is the length of the data you want to send (in bytes).
- flags is set to 0.
- to is a pointer to struct sockaddr for the host where data has to be sent.
- tolen is set it to sizeof(struct sockaddr).

The recvfrom Function:

The *recvfrom* function is used to receive data from UNCONNECTED datagram sockets. Put simply, when you use scoket type as SOCK_DGRAM

```
int recvfrom(int sockfd, void *buf, int len, unsigned int flags
                    struct sockaddr *from, int *fromlen);
```

This call returns the number of bytes read into the buffer otherwise it will return -1 on error.

Parameters:

- sockfd: is a socket descriptor returned by the socket function.
- buf is the buffer to read the information into.
- len is the maximum length of the buffer.
- flags is set to 0.
- from is a pointer to struct sockaddr for the host where data has to be read.
- fromlen is set it to sizeof(struct sockaddr).

The close Function:

The *close* function is used to close the communication between client and server.

```
int close( int sockfd );
```

This call returns 0 on success otherwise it will return -1 on error. Parameters:

- sockfd: is a socket descriptor returned by the socket function.

The *shutdown* Function:

The *shutdown* function is used to gracefully close the communication between client and server. Thisfunction gives more control in caomparision of *close* function.

```
int shutdown(int sockfd, int how);
```

This call returns 0 on success otherwise it will return -1 on error.

Parameters:

- sockfd: is a socket descriptor returned by the socket function.
- how: put one of the numbers:
  - 0 indicates receives disallowed,
  - 1 indicatesthat sends disallowed and
  - 2 indicates that sends and receives disallowed. When how is set to 2, it's the same thingas close().

The select Function:

The *select* function indicates which of the specified file descriptors is ready for reading, ready for writing,or has an error condition pending.

When an application calls *recv or recvfrom* it is blocked until data arrives for that socket. An application could be doing other useful processing while the incoming data stream is empty. Another situation is when an application receives data from multiple sockets.

Calling *recv or recvfrom* on a socket that has no data in it's input queue prevents immediate reception of data from other sockets. The select function call solves this problem by allowing the program to poll all the socket handles to see if they are available for non-blocking reading and writing operations.

```
int select(int nfds, fd_set *readfds, fd_set *writefds,
   fd_set *errorfds, struct timeval *timeout);
```

This call returns 0 on success otherwise it will return -1 on error.

Parameters:

- nfds: specifies the range of file descriptors to be tested. The select() function tests file descriptors in the range of 0 to nfds-1
- readfds:points to an object of type *fd_set* that on input specifies the file descriptors to be checked for being ready to read, and on output indicates which file descriptors are ready to read. Can be NULL to indicate an empty set.
- writefds:points to an object of type *fd_set* that on input specifies the file descriptors to be checkedfor being ready to write, and on output indicates which file descriptors are ready to write Can be NULL to indicate an empty set.
- exceptfds :points to an object of type *fd_set* that on input specifies the file descriptors to be checked for error conditions pending, and on output indicates which file descriptors have

error conditions pending. Can be NULL to indicate an empty set.

- timeout :poins to a timeval struct that specifies how long the select call should poll the descriptors for an available I/O operation. If the timeout value is 0, then select will return immediately. If the timeout argument is NULL, then select will block until at least one file/socket handle is ready for an available I/O operation. Otherwise select will return after the amount of time in the timeout has elapsed OR when at least one file/socket descriptor is ready for an I/O operation.

The return value from select is the number of handles specified in the file descriptor sets that are ready forI/O. If the time limit specified by the timeout field is reached, select return 0. The following macros exist for manipulating a file descriptor set:

- FD_CLR(fd, &fdset): Clears the bit for the file descriptor fd in the file descriptor set *fdset*
- FD_ISSET(fd, &fdset): Returns a non-zero value if the bit for the file descriptor *fd* is set in thefile descriptor set pointed to by *fdset*, and 0 otherwise.
- FD_SET(fd, &fdset): Sets the bit for the file descriptor fd in the file descriptor set fdset.
- FD_ZERO(&fdset): Initializes the file descriptor set fdset to have zero bits for all file descriptors.

The behavior of these macros is undefined if the fd argument is less than 0 or greater than or equal to FD_SETSIZE.

Example:

```
fd_set fds;

struct timeval tv;

/* do socket initialization etc. tv.tv_sec = 1;

tv.tv_usec = 500000;

/* tv now represents 1.5 seconds */ FD_ZERO(&fds);

/* adds sock to the file descriptor set */ FD_SET(sock, &fds);
```

```
/* wait 1.5 seconds for any data to be read from any single socket */
select(sock+1, &fds, NULL, NULL, &tv); if (FD_ISSET(sock, &fds))
{
recvfrom(s, buffer, buffer_len, 0, &sa, &sa_len);
/* do something */
}
else
{
/* do something else */
}
```

There are various structures which are used in Unix Socket Programming to hold information about the address and port and other information. Most socket functions require a pointer to a socket address structure as an argument. Structures defined in this tutorial are related to Internet Protocol Family.

The first structure is struct *sockaddr* that holds socket information:

```
struct sockaddr{
        unsigned short sa_family;
        char           sa_data[14];
};
```

This is a generic socket address structure which will be passed in most of the socket function calls. Here is the description of the member fields:

Attribute	Values	Description
sa_family	AF_INET AF_UNIX AF_NS AF_IMPLINK	This represents an address family. In most of the Internet basedapplications we use AF_INET.
sa_data	Protocol SpecificAddress	The content of the 14 bytes of protocol specific address are interpreted according to the type of address. For the Internet family we will use port number IP address which is represented by *sockaddr_in* structure definedbelow.

Second structure that helps you to reference to the socket's elements is as follows:

```
struct in_addr {

        unsigned long s_addr;

};
```

Here is the description of the member fields:

Attribute	Values	Description
sa_family	AF_INET AF_ UNIX AF_NS AF_IMPLINK	This represents an address family. In most of the Internet based applicationswe use AF_ INET.
sin_port	Service Port	A 16 bit port number in Network Byte Order.
sin_addr	IP Address	A 32 bit IP address in Network Byte Order.
sin_zero	Not Used	You just set this value to NULL as this is not being used.

The next structure is used only in the above structure as a structure field and holds 32 but netid/hostid.

```
struct hostent

{

  char *h_name;

  char **h_aliases;

  int  h_addrtype;
```

Here is the description of the member fields:

Attribute	Values	Description
s_addr	service port	A 32 bit IP address in Network Byte Order.

There is one more important structure. This structure is used to keep information related to host.

Here is the description of the member fields:

Attribute	Values	Description
h_name	ti.com etc	This is official name of the host. For example tutorial-spoint.com, google.cometc.
h_aliases	TI	This will hold a list of host name aliases.
h_addr-type	AF_ INET	This contains the address family and in case of Internet based application it willalways be AF_INET
h_length	4	This will hold the length of IP address which is 4 for Internet Address.
h_addr_ list	in_addr	For the Internet addresses the array of pointers h_addr_list[0], h_addr_list[1] andso on are points to structure in_addr.

NOTE: h_addr is defined as h_addr_list[0] to keep backward compatibility. Following structure is used to keep information related to service and associated ports.

```
struct servent
{
  char *s_name;
  char **s_aliases;
  int  s_port;
  char  *s_proto;
};
```

Here is the description of the member fields:

Attribute	Values	Description
s_name	http	This is official name of the service. For example SMTP, FTP POP3 etc.
s_aliases	ALIAS	This will hold list of service aliases. Most of the time this will be set to NULL.
s_port	80	This will have associated port number. For example for HTTP this will be 80.
s_proto	TCP UDP	This will be set to the protocol used. Internet services are provided using either TCPor UDP.

Additional Topics

1) Pipes:

```
#include <unistd.h>
#include <stdlib.h>
#include <stdio.h>
int main()
{
	FILE *write_fp;
	char buffer[BUFSIZ + 1];
	sprintf(buffer, "Once upon a time, there was ...\n");write_fp
	= popen ("od -c", "w");
	if (write_fp != NULL){
		fwrite( buffer , sizeof(char), strlen(buffer), write_fp);
		pclose(write_fp);
		exit(EXIT_SUCCESS);
	}
	exit(EXIT_FAILURE);
```

Reading Output From an External Program:

```
#include <unistd.h>
#include <stdlib.h>
#include <stdio.h>
#include <string.h>
int main()
{
	FILE *read_fp;
	char buffer[BUFSIZ + 1];
	int chars_read;
	memset(buffer,'\0',sizeof(buffer));
	read_fp = popen("uname -a", "r");
	if (read_fp !=NULL){
		chars_read = fread(buffer, sizeof(char), BUFSIZ, read_fp);
		if (chars_read > 0){
			printf("Output was:-\n%s\n",buffer);
		}
		pclose(read_fp);
```

```
        exit(EXIT_SUCCESS);
    }
    exit(EXIT_FAILURE);
```

2) Creating a Named Pipe:

```
#include <unistd.h>
#include <stdlib.h>
#include <stdio.h>
#incldue <sys/types.h>
#include <sys/stat.h>
Int main()
{
    int res = mkfifo("/tmp/my_fifo",0777);
    if (res == 0) printf("FIFO created \n");
    exit (EXIT_SUCCESS);
}
```

3) Inter-process Communication with FIFOs:

```
#include <unistd.h>
#include <stdlib.h>
#include <stdio.h>
#include <string.h>
#include <fcntl.h>
#include <limits.h>
#include <sys/types.h>
#include <sys/stat.h>
#define FIFO_NAME "/tmp/my_file"
#define BUFFER_SIZE PIPE_BUF
#define TEN_MEG (1024*1024*10)
Int main()
{
    int pipe_fd;int res;
    int open_mode = O_WRONLY;
    int bytes_sent = 0;
    char buffer[BUFFER_SIZE + 1];
    if (access(FIFO_NAME, F_OK) == -1){
        res = mkfifo(FIFO_NAME, 0777);
        if (res != 0){
```

```
            fprintf(stderr, "Could not create fifo %s\n", FIFO_NAME);
           exit(EXIT_FAILURE)
            }
    }
    printf("Process %d opening FIFO O_WRONLY\n",getpid());
    pipe_fd = open(FIFO_NAME, open_mode);
    printf("Process %d result %d\n", getpid(),pipe_fd);
}
```

■■■

Made in United States
Troutdale, OR
11/03/2026

56845392R00149